Deno Web Development

Write, test, maintain, and deploy JavaScript and
TypeScript web applications using Deno

Alexandre Portela dos Santos

BIRMINGHAM—MUMBAI

Deno Web Development

Associate Group Product Manager: Pavan Ramchandani

Publishing Product Manager: Aaron Tanna

Commissioning Editor: Pavan Ramchandani

Senior Editor: Sofi Rogers

Content Development Editor: Rakhi Patel

Technical Editor: Saurabh Kadave

Copy Editor: Safis Editing

Project Coordinator: Manthan Patel

Proofreader: Safis Editing

Indexer: Rekha Nair

Production Designer: Prashant Ghare

First published: March 2021

Production reference: 1240321

Published by Packt Publishing Ltd.

Livery Place

35 Livery Street

Birmingham

B3 2PB, UK.

ISBN 978-1-80020-566-6

www.packt.com

To my parents, Fátima and António, and to my brother Pedro. Thanks for your unconditional support – you are the best. Wouldn't have made it without you.

To my friends, who helped me make this book a reality. Thank you, Felipe, Gonçalo, Bruno, Nuno, János, João, and Miguel for all the conversations, reviews, and suggestions.

– Alexandre

Foreword

Many of us witnessed a time where JavaScript was not considered a real language but rather a scripting tool for those who wanted to improve the user experience when interacting with the browser.

Its increasing popularity drove the JavaScript community to broaden its purpose. Suddenly, someone who was mainly skilled in client-side web development now had the chance to dive into the world of server-side with no language barrier. On top of that, they had something quite natural that was becoming trendy – concurrency and event-driven design.

At the same time, looking around the programming language landscape over the last 10-15 years, other languages were being released as well, positioned to be more advanced alternatives leveraging the learnings of the traditional ones; simpler to package; removing legacy; providing more advanced standard libraries and features so that we developers could focus on delivering value to the end users of the software. There are, however, two trends worth highlighting. Concurrency and interoperability were topics we could see being addressed in all of them. Golang, Kotlin, Elixir, and Rust offered easier ways to handle concurrency, but they also provided an easy way to interop with other languages. You could use Erlang code in Elixir, Java code in Kotlin, C in Golang and Rust, and so on... Interop was critical for language adoption in existing legacy systems.

On one hand, most of these languages did well in solving a problem that Node.js was also solving while being quite mature because they leveraged existing ecosystems with interop capabilities. On the other hand, Node.js suffers from some design decisions that drag improvement in different areas such as security, packaging, and interoperability to name a few. One might say there's some catching up to be done...

Being a software engineer means being in an infinite loop of learning. This book is a remarkable achievement and a must-read in one iteration of that learning loop. It is an introduction to a new language that promises to fill some gaps, building a strong foundation to consolidate the reason why the most popular language in the world is even more everywhere – *Deno Web Development: Write, test, maintain, and deploy JavaScript and TypeScript web applications using Deno.*

Miguel Loureiro

Chief Technology Officer, KI group

Contributors

About the author

Alexandre Portela dos Santos is a software engineer passionate about products and start-ups. For the last 8+ years, he's been working together with multiple companies, using technology as an enabler for ideas and businesses. With a big interest in education and getting people excited about technology, he makes sure he's always involved with people that are learning about it, be it via blog posts, books, open source contributions, or meetups. This is, by itself, a learning adventure that Alexandre loves to be a part of. Being a true believer that great software only happens through collaboration, ownership, and teams of great people, he strives to nurture those values in every project he works on.

About the reviewers

Maxim Mazurok is a software engineer with a focus on web development. He has 6+ years of enterprise experience in software product and e-commerce companies located in Silicon Valley, Ukraine, and Sydney.

He has applied this experience to teach teens the basics of web development and taught adults more advanced web technologies.

He is passionate about open source and sharing knowledge, an active GitHub contributor, and a Stack Overflow member, always happy to connect.

When not at the computer, he rides a bicycle, takes pictures of nature, and attends meetups and conferences.

Yusuke Tanaka is a software engineer living in Tokyo, Japan. After getting a BS degree in engineering at the University of Tokyo in 2019, he now works at STADIUM Co., Ltd. engaging in the development of a web application that provides an online job interview system. Although he uses TypeScript and Go at work, he has been enthusiastic about the Rust programming language for about 2 years and makes contributions to open source software related to Rust in his spare time. In particular, he is so interested in Deno and the core of Rust that he devotes a lot of time to them. To get in touch with him or for more details, feel free to visit his GitHub (@magurotuna) or Twitter (@yusuktan).

Acknowledgement

Firstly, I am grateful to the editors and publishers who gave me a chance to participate in such a wonderful project. And most of all, I would like to thank Alexandre Santos for making this book happen. I am delighted to get involved in this book, hoping it will help more people recognize the fantastic development experience with Deno.

Table of Contents

3

The Runtime and Standard Library

Section 2: Building an Application

4

Building a Web Application

Section 3: Testing and Deploying

8

Testing – Unit and Integration

9

Deploying a Deno Application

10

What's Next?

Other Books You May Enjoy

Index

Preface

Deno is a JavaScript/TypeScript runtime with secure defaults and a great developer experience.

Deno Web Development introduces Deno's primitives, its principles, and how developers can use them to build real-world applications. The book is divided into three main sections: introducing Deno, building an API from scratch, and testing and deploying a Deno application. By the end of the book, the reader will be comfortable in using Deno to create, maintain, and deploy secure and reliable web applications.

Who this book is for

This book targets developers of all levels who want to leverage their JavaScript and TypeScript skills in a secure, simple, and modern runtime, using it for web development.

What this book covers

Chapter 1, What is Deno?, gives historical context about Node.js and the motivations that led to Deno's creation, presenting the runtime architecture and premises.

Chapter 2, The Toolchain, covers installing Deno and explores the tools included in the runtime binary.

Chapter 3, The Runtime and Standard Library, explains about writing simple scripts and applications using Deno's runtime and standard library functions.

Chapter 4, Building a Web Application, shows how to set up the foundations for a web application using the standard library HTTP module.

Chapter 5, Adding Users and Migrating to Oak, covers using oak, a popular HTTP library, to build a REST API, and adding persistence and users to the application.

Chapter 6, Adding Authentication and Connecting to the Database, looks at adding support for authentication and authenticated endpoints, and connecting to a MongoDB database.

Chapter 7, HTTPS, Extracting Configuration, and Deno in the Browser, looks at enabling HTTP, handling configuration based on files and the environment, and Deno code in the browser.

Chapter 8, Testing – Unit and Integration, covers writing and running unit and integration tests for the modules written in previous chapters.

Chapter 9, Deploying a Deno Application, goes into configuring a container environment and automation to deploy a Deno application, getting it running in a cloud environment.

Chapter 10, What's Next?, gives an overview of what we have learned throughout the book, a roadmap of Deno, explains how to publish a module to Deno's official registry, and talks you through the future and community of Deno.

To get the most out of this book

All the code examples in this book were tested using Deno 1.7.5 on macOS, but they should work in future releases of Deno. A few third-party packages were also used in the course of the book. The examples using them should also apply to newer versions of the software.

The book will provide installation instructions for all the pieces of software used.

Software/Hardware covered in the book	OS Requirements
Deno 1.7.5	Windows, macOS, and Linux

This book's code was written using VS Code (https://code.visualstudio.com/) for the best experience using the official Deno extension. This is not a requirement and the book can be followed using any coding editor.

If you are using the digital version of this book, we advise you to type the code yourself or access the code via the GitHub repository (link available in the next section). Doing so will help you avoid any potential errors related to the copying and pasting of code.

You should be comfortable using JavaScript and have basic knowledge of TypeScript. Node.js knowledge is not required but might be useful.

Download the example code files

You can download the example code files for this book from your account at www.packt.com. If you purchased this book elsewhere, you can visit www.packtpub.com/support and register to have the files emailed directly to you.

You can download the code files by following these steps:

1. Log in or register at www.packt.com.

2. Select the **Support** tab.

3. Click on **Code Downloads**.

4. Enter the name of the book in the **Search** box and follow the onscreen instructions.

Once the file is downloaded, please make sure that you unzip or extract the folder using the latest version of:

- WinRAR/7-Zip for Windows

- Zipeg/iZip/UnRarX for Mac

- 7-Zip/PeaZip for Linux

The code bundle for the book is also hosted on GitHub at https://github.com/PacktPublishing/Deno-Web-Development. In case there's an update to the code, it will be updated on the existing GitHub repository.

We also have other code bundles from our rich catalog of books and videos available at https://github.com/PacktPublishing/. Check them out!

Conventions used

There are a number of text conventions used throughout this book.

Code in text: Indicates code words in text, database table names, folder names, filenames, file extensions, pathnames, dummy URLs, user input, and Twitter handles. Here is an example: "Add oak-middleware-jwt to the deps.ts file and export the jwtMiddleware function."

A block of code is set as follows:

```
const apiRouter = new Router({ prefix: "/api" })
apiRouter.use(async (_, next) => {
  console.log("Request was made to API Router");
  await next();
}))
...
app.use(apiRouter.routes());
app.use(apiRouter.allowedMethods());
```

When we wish to draw your attention to a particular part of a code block, the relevant lines or items are set in bold:

```
const app = new Application();
app.use(async (ctx, next) => {
  const start = Date.now();
  await next();
  const ms = Date.now() - start;
  ctx.response.headers.set("X-Response-Time", `${ms}ms`);
});
...
app.use(apiRouter.routes());
app.use(apiRouter.allowedMethods());
```

Any command-line input or output is written as follows:

```
$ deno --version

deno 1.7.5 (release, x86_64-apple-darwin)
v8 9.0.123
typescript 4.1.4
```

Bold: Indicates a new term, an important word, or words that you see onscreen. For example, words in menus or dialog boxes appear in the text like this. Here is an example: "If you've used MongoDB, you can see your users created there on the Atlas interface, by going to the **Collections** menu."

> **Tips or important notes**
> Appear like this.

Get in touch

Feedback from our readers is always welcome.

General feedback: If you have questions about any aspect of this book, mention the book title in the subject of your message and email us at customercare@packtpub.com.

Errata: Although we have taken every care to ensure the accuracy of our content, mistakes do happen. If you have found a mistake in this book, we would be grateful if you would report this to us. Please visit www.packtpub.com/support/errata, selecting your book, clicking on the Errata Submission Form link, and entering the details.

Piracy: If you come across any illegal copies of our works in any form on the Internet, we would be grateful if you would provide us with the location address or website name. Please contact us at copyright@packt.com with a link to the material.

If you are interested in becoming an author: If there is a topic that you have expertise in and you are interested in either writing or contributing to a book, please visit authors.packtpub.com.

Reviews

Please leave a review. Once you have read and used this book, why not leave a review on the site that you purchased it from? Potential readers can then see and use your unbiased opinion to make purchase decisions, we at Packt can understand what you think about our products, and our authors can see your feedback on their book. Thank you!

For more information about Packt, please visit packt.com.

Section 1: Getting Familiar with Deno

In this section, you will get to know what Deno is, why it was created, and how it was created. This section will help you set up the environment and get familiar with the ecosystem and available tooling.

This section contains the following chapters:

- *Chapter 1, What Is Deno?*
- *Chapter 2, The Toolchain*
- *Chapter 3, The Runtime and Standard Library*

1
What is Deno?

Deno is a secure runtime for JavaScript and TypeScript. I'll guess that you are probably getting that excitement of experimenting with a new tool. You have worked with JavaScript or TypeScript and have at least heard of Node.js. Deno will feel like it has the right amount of novelty for you and, at the same time, has some things that will sound familiar for someone working in the ecosystem.

Before we start getting our hands dirty, we'll understand how Deno was created and its motivations. Doing that will help us learn and understand it better.

We'll be focusing on practical examples throughout this book. We'll be writing code and then rationalizing and explaining the underlying decisions we've made. If you come from a Node.js background, some of the concepts might sound familiar to you. We will also explain Deno and compare it with its ancestor, Node.js.

Once the fundamentals are in place, we'll dive into Deno and explore its runtime features by building small utilities and real-world applications.

Without Node, there would be no Deno. To understand the latter well, we can't ignore its 10+ year-old ancestor, which is what we'll look at in this chapter. We'll explain the reasons for its creation back in 2009 and the pain points that were detected after a decade of usage.

After that, we'll present Deno and the fundamental differences and challenges it proposes to solve. We'll have a look at its architecture, some principles and influences of the runtime, and the use cases where it shines.

After understanding how Deno came to life, we will explore its ecosystem, its standard library, and some use cases where Deno is instrumental.

Once you've read this chapter, you'll be aware of what Deno is and what it is not, why it is not the next version of Node.js, and what to think about when you're considering Deno for your next project.

In this chapter, we'll cover the following topics:

- A little history
- Why Deno?
- Architecture and technologies that support Deno
- Grasping Deno's limitations
- Exploring Deno's use cases

Let's get started!

A little history

Deno's first stable version, v1.0.0, was launched on the May 13, 2020.

The first time Ryan Dahl – Node.js creator – mentioned it was in his famous talk, *10 things I regret about node.js* (`https://youtu.be/M3BM9TB-8yA`). Apart from the fact that it presents the first very alpha version of Deno, it is a talk worth watching as a lesson on how software ages. It is an excellent reflection on how decisions evolve, even when they're made by some of the smartest people in the open source community, and how they can end up in a different place than what they initially planned for.

After the launch, in May 2020 and due to its historical background, its core team, and the fact that it appeals to the JavaScript community, Deno has been getting lots of attention. That's probably one way you've heard about it, be it via blog posts, tweets, or conference talks.

This enthusiasm is having positive consequences on its runtime, with lots of people wanting to contribute and use it. The community is growing due to its Discord channel (`https://discord.gg/deno`) and the number of pull requests on Deno's repositories (`https://github.com/denoland`). It is currently evolving at a cadence of one minor version per month, with lots of bug fixes and improvements being shipped. The roadmap shows a vision for a future that is no less exciting than the present. With a well-defined path and set of principles, Deno has everything it takes to become more significant by the day.

Let's rewind a little and go back to 2009 and the creation of Node.js.

At the time, Ryan started by questioning how most backend languages and frameworks were dealing with I/O (input/output). Most of the tools were looking at I/O as an synchronous operation, blocking the process until it is done, and only then continuing to execute the code.

Fundamentally, it was this synchronous blocking operation that Ryan questioned.

Handling I/O

When you are writing servers that must deal with thousands of requests per second, resource consumption and speed are two significant factors.

For such resource-critical projects, it is important that the base tools – the primitives – have an architecture that is accounting for this. When the time to scale arises, it helps that the fundamental decisions you made at the beginning support that.

Web servers are one of those cases. The web is a significant platform in today's world. It never stops growing, with more devices and new tools accessing the internet daily, making it accessible to more people. The web is the common, democratized, decentralized ground for people around the world. With this in mind, the servers behind those applications and websites need to handle giant loads. Web applications such as Twitter, Facebook, and Reddit, among many others, deal with thousands of requests per minute. So, scale is essential.

To kickstart a conversation about performance and resource efficiency, let's look at the following graph, which is comparing two of the most used open-source web servers: Apache and Nginx:

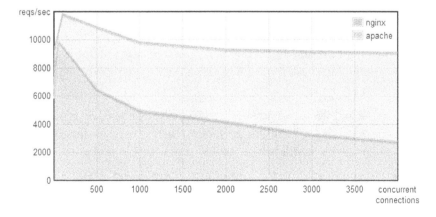

Figure 1.1 – Requests per second versus concurrent connections – Nginx versus Apache

At first glance, this tells us that Nginx comes out on top pretty much every time. We can also understand that, as the number of concurrent connections increases, Apache's number of requests per second decreases. Comparatively, Nginx keeps the number of requests per second pretty stable, despite also showing an expected drop in requests per second as the number of connections grows. After reaching a thousand concurrent connections, Nginx gets close to double the number of Apache's requests per second.

Let's look at a comparison of the RAM memory consumption:

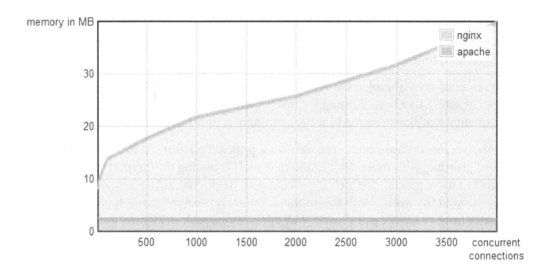

Figure 1.2 – Memory consumption versus concurrent connections – Nginx versus Apache

Apache's memory consumption grows *linearly* with the number of concurrent connections, while Nginx's memory footprint is constant.

You might already be wondering why this happens.

This happens because Apache and Nginx have very different ways of dealing with concurrent connections. Apache spawns a new thread per request, while Nginx uses an event loop.

In a *thread-per-request* architecture, it creates a thread every time a new request comes in. That thread is responsible for handling the request until it finishes. If another request comes while the previous one is still being handled, a new thread is created.

On top of this, handling networking on threaded environments is not known as something particularly easy to do. You can incur in file and resource locking, thread communication issues, and common problems such as deadlocks. Adding to the difficulties presented to the developer, using threads does not come for free, as threads by themselves have a resource overhead.

In contrast, in an event loop architecture, everything happens on a single thread. This decision dramatically simplifies the lives of developers. You do not have to account for the factors mentioned previously, which means more time to deal with your users' problems.

By using this pattern, the web server just sends events to the event loop. It is an asynchronous queue that executes operations when there are available resources, returning to the code asynchronously when these operations finish. For this to work, all the operations need to be non-blocking, meaning they shouldn't wait for completion and just send an event and wait for a response later.

Blocking versus non-blocking

Take, for instance, reading a file. In a blocking environment, you would read the file and have the process waiting for it to finish until you execute the next line of code. While the operating system is reading the file's contents, the program is in an idle state, wasting valuable CPU cycles:

```
const result = readFile('./README.md');
// Use result
```

The program will wait for the file to be read and only then it will continue executing the code.

The same operation using an event loop would be to trigger the "read the file" event and execute other tasks (for instance, handling other requests). When the file reading operation finishes, the event loop will call the callback function with the result. This time, the runtime uses the CPU cycles to handle other requests while the OS retrieves the file's contents, making better use of the resources:

```
const result = readFileAsync('./README.md', function(result) {
    // Use result
});
```

In this example, the task gets a callback assigned to it. When the job is complete (this might take seconds or milliseconds), it calls back the function with the result. When this function is called, the code inside runs linearly.

Why aren't event loops used more often?

Now that we understand the advantages of event loops, this is a very plausible question. One of the reasons event loops are not used more, even though there are some implementations in Python and Ruby, is that they require all the infrastructure and code to be non-blocking. Being non-blocking means being prepared not to execute the code synchronously. It means triggering events and dealing with the result later, at some point in time.

On top of all of that, many of the commonly used languages and libraries do not provide asynchronous APIs. Callbacks are not present in many languages, and anonymous functions do not exist in programming languages such as C. Crucial pieces of today's software, such as `libmysqlclient`, do not support asynchronous operations, even though part of its internals might use asynchronous task execution. Asynchronous DNS resolution is another example that's also not a standard in many systems. As another example, you might take, for instance, the manual pages of operating systems. Most of them don't even provide us with a way to understand if a particular function does I/O or not. These are all evidences that the ability to make asynchronous I/O is not present in many of today's fundamental software pieces.

Even the existing tools that provide these features require developers to have a deep understanding of asynchronous I/O patterns to use event loops. It's a difficult job to wire up these existing solutions to get something to work while going around technical limitations, such as the ones shown in the `libmysqlclient` example.

JavaScript to the rescue

JavaScript was a language created by Brendan Eich in 1995 while working for Netscape. It initially only ran in browsers and allowed developers to add interactive features to web pages. It is composed of elements that revealed themselves as perfect for the event loop:

- It has anonymous functions and closures.
- It only executes one callback at a time.
- I/O is done on DOM via callbacks (for example, `addEventListener`).

Combining these three fundamental aspects of the language made the event loop something natural to anyone used to JavaScript in the browser.

The language's features ended up gearing its developers toward event-driven programming.

Node.js enters the scene

After all these thoughts and questions about I/O and how it should be dealt with, Ryan Dahl came up with Node.js in 2009. It is a JavaScript runtime, based on Google's V8 – a JavaScript engine that brings JavaScript to the server.

Node.js is asynchronous and single-threaded by design. It has an event loop at its core and presents itself as a scalable way to develop backend applications that handle thousands of concurrent requests.

Event loops provide us with a clean way to deal with concurrency, a topic where Node.js contrasts with tools such as PHP or Ruby, which use the thread-per-request model. This single-threaded environment grants Node.js users the simplicity of not caring about thread-safety problems. It very much succeeds in abstracting the event loop and all the issues with synchronous tools from the user, requiring little to no knowledge about the event loop itself. Node.js does this by leveraging callbacks and, more recently, the use of promises.

Node.js positioned itself as a way to provide a low-level, purely evented, non-blocking infrastructure for users to program their applications.

Node.js' rise

Telling companies and developers that they could leverage their JavaScript knowledge to write servers rapidly resulted in a Node.js popularity rise.

It didn't take much time for the language to evolve fast since it was released and started being used in production by companies of all sizes.

Just 2 years after its creation, in 2011, Uber and LinkedIn were already running JavaScript on the server. In 2012, Ryan Dahl resigned from the Node.js community's day-to-day operations to dedicate himself to research and other projects.

Estimates say that, in 2017, there were more than 8.8 million instances of Node.js running (source: `https://blog.risingstack.com/history-of-node-js/`). Today, more than 103 billion packages have been downloaded from **Node Package Manager** (**npm**), and there are around 1,467,527 packages published.

Node.js is a great platform, there's no questions about that. Pretty much anyone who has used it has experienced many of its advantages. Popularity and community play a significant role in this. Having a lot of people of very different experience levels and backgrounds working with a piece of technology can only push it forward. That's what happened – and still happens – with Node.js.

Node.js enabled developers to use JavaScript for lots of varying use cases that weren't possible previously. This ranged from robotics, to cryptocurrencies, to code bundlers, APIs, and more. It is a stable environment where developers feel productive and fast. It will continue its job, supporting companies and businesses of different sizes for many years to come.

But you've bought this book, so you must believe that Deno has something worth exploring, and I can guarantee that it does.

You might be wondering, why bring a new solution to the table when the previous one is more than satisfactory? That's what we'll discover next.

Why Deno?

Many things have changed since Node.js was created. More than a decade has passed; JavaScript has evolved, as well as the software infrastructure community. Languages such as Rust and golang were born and were very important developments in the software community. These languages made it much easier to produce native machine code while providing a strict and reliable environment for developers to work on.

However, this strictness comes at the cost of productivity. Not that developers don't feel productive writing those languages, because they do, but you can easily argue that productivity is a subject where dynamic languages clearly shine.

The ease and speed of developing dynamic languages makes them a very strong contestant when it comes to scripting and prototyping. And when it comes to thinking about dynamic languages, JavaScript directly comes to mind.

JavaScript is the most used dynamic language and runs in every device with a web browser. Due to its heavy usage and giant community, many efforts have been put into optimizing it. The creation of organizations such as ECMA International has ensured that the language evolves stably and carefully.

As we saw in the previous section, Node.js played a very successful role in bringing JavaScript to the server, opening the door to a huge amount of different use cases. It's currently used for many different tasks, including web development tooling, creating web servers, and scripting, among many others. At the time of its creation, and to enable such use cases, Node.js had to invent concepts for JavaScript that didn't exist before. Later, these concepts were discusses by the standards' organizations and added to the language differently, making parts of Node.js incompatible with its mother language, ECMAScript. A decade has passed, and ECMAScript has evolved, as well as the ecosystem around it.

CommonJS modules are no longer the standard; JavaScript has ES modules now. *TypedArrays* are now a thing, and finally, JavaScript can directly handle binary data. Promises and async/await are the go-to way with asynchronous operations.

These features are available on Node.js, but they must coexist with the non-standard features that were created back in 2009 that still need to be maintained. These features, and the large number of users that Node.js has, made it difficult and slow to evolve the system.

To solve some of these problems, and to keep up with the evolution of the JavaScript language, many community projects were created. These projects made it possible for us to use the latest features of the language but added things such as a build system to many Node.js projects, heavily complicating them. Quoting Dahl, it "*took away from the fun of dynamic language scripting.*"

More than 10 years of heavy usage also made it clear that some of the runtime's fundamental constructs needed improvement. A lack of a security sandbox was one of the major issues. At the time Node.js was created, it was possible for JavaScript to access the "outside world" by creating bindings in V8 – the JavaScript engine behind it. Even though these bindings enabled I/O features such as reading from the filesystem accessing the network, they also broke the purpose of the JavaScript sandbox. This decision made it really hard to let the developer control what a Node.js script has access to. In its current state, for instance, there's nothing preventing a third-party package in a Node.js script to read all the files the user has access to, among performing other nefarious actions.

A decade later, Ryan Dahl and the team behind Deno were missing a fun and productive scripting environment that could be used for a wide range of tasks. The team also felt like the JavaScript landscape has changed enough that it was worthwhile simplifying, and thus they decided to create Deno.

Presenting Deno

"Deno is a simple, modern, and secure runtime for JavaScript and TypeScript that uses V8 and is built into Rust."
– `https://deno.land/`

Deno's name was constructed by inverting the syllables of its ancestor's name, no-de, de-no. With a lot of lessons learned from its ancestor, Deno presents the following as its main features:

- Secure by default
- First-class TypeScript support

- A single executable file

- Provides fundamental tools to write applications

- Complete and audited standard library

- Compatibility with ECMAScript and browser environments

Deno is secure by default, and it was created like that by design. It ultimately leverages the V8 sandbox and provides a strict permission model that enables developers to finely control what the code has access to.

TypeScript is also first-class supported, meaning developers can choose to use TypeScript without any extra configuration. All the Deno APIs are also written in TypeScript and thus have correct and precise types and documentation. The same is true for the standard library.

Deno ships a single executable with all the fundamental tools needed to write applications; it will always be that way. The team makes an effort to keep the executable small (~15 MB) so that we can use it in various situations and environments, from simple scripts to full-fledged applications.

More than just executing code, the Deno binary provides a complete set of developer utilities, namely a linter, a formatter, and a test runner.

Golang's carefully polished standard library inspired Deno's standard library. It is deliberately bigger and more complete compared to Node.js'. This decision was made to address the enormous dependency trees that used to occur in some Node.js projects. Deno's core team believes that, by providing a stable and complete standard library, it can help address this problem. By removing the need to create third-party packages to handle common use cases the platform provides by default, it aims to diminish the need to use loads of third-party packages.

To keep compatibility with ES6 and browsers, Deno made efforts to mimic browser APIs. Things such as performing HTTP requests, dealing with URLs, or encoding text, among others, can be done by using the same APIs you'd use in a browser. A deliberate effort was made by the Deno team to keep these APIs in sync with the browser.

Aiming to offer the best of three worlds, Deno provides the prototype-ability and developer experience of JavaScript, the type-safety and security offered by Typescript, and Rust's performance and simplicity.

Ideally, as Dahl also mentioned in one of his talks, code would follow the following flow in the path from prototype to production: developers can start writing JavaScript, migrate to TypeScript, and end up with Rust code.

At the time of writing, is it only possible to run JavaScript and TypeScript. Rust is only available via a (still unstable) plugin API that might become stable in a not-so-distant future.

A web browser for command-line scripts

As time passed, the Node.js module system evolved into something that is now overly complex and painful to maintain. It takes into consideration edge cases such as importing folders, searching for dependencies, importing relative files, searching for index.js, third-party packages, and reading the `package.json` file, among others.

It also got heavily coupled with **npm**, the **Node Package Manager**, which was initially part of Node.js itself but separated in 2014.

Having a centralized package manager is not very webby, to use Dahl's words. The fact that millions of applications depend on a single registry to survive is a liability.

Deno solves this problem by using URLs. It takes an approach that's very similar to a browser, only requiring an absolute URL to a file to execute or import code. This absolute URL can be local, remote, or HTTP-based and includes the following file extension:

```
import { serve } from 'https://deno.land/std@0.83.0/http/
server.ts'
```

The preceding code happens to be the same code you would write on a browser inside a `<script>` tag if you want to require an ES module.

In regard to installation and offline usage, Deno ensures that users don't have to worry about that by using a local cache. When the program runs, it installs all the required dependencies, removing the need for an installation step. We'll dive into this later in more depth later, in *Chapter 2, The Toolchain*.

Now that we are comfortable with what Deno is and the problems it solves, we're in good shape to go beyond the surface. By knowing what is happening behind the scenes, we can get a better comprehension of Deno itself.

In the next section, we'll explore technologies that support Deno and how they connect.

Architecture and technologies that support Deno

Architecture-wise, Deno took various topics into consideration such as security. Deno put much thought into establishing a clean and performant way of communicating with the underlying OS without leaking details to the JavaScript side. To enable that, Deno uses message-passing to communicate from inside the V8 to the Deno backend. The backend is the component written in Rust that interacts with the event loop and thus with the OS.

Deno has been made possible by four pieces of technology:

- V8

- TypeScript

- Tokio (event loop)

- Rust

It is the connection of all those four parts that make it possible to provide developers with a great experience and development speed while keeping the code safe and sandboxed. If you are not familiar with these pieces of technology, I'll leave a short definition:

V8 is a JavaScript engine developed by Google. It is written in C++ and runs across all major operating systems. It is also the engine behind Chrome, Node.js, and others.

TypeScript is a superset of JavaScript developed by Microsoft that adds optional static typing to the language and *transpiles* it to JavaScript.

Tokio is an asynchronous runtime for Rust that provides utilities to write network applications of any scale.

Rust is a server-side language designed by Mozilla focused on performance and safety.

Using Rust, a fast-growing language, to write Deno's core made it more approachable for developers than Node.js. Node.js' core was written in C++, which is not known for being exceptionally easy to deal with. With many pitfalls and with a not-so-good developer experience, C++ revealed itself as a small obstacle in the evolution of Node.js core.

`Deno_core` is shipped as a Rust crate (package). This connection with Rust is not a coincidence. Rust provides many features that facilitate this connection with JavaScript and adds capabilities to Deno itself. Asynchronous operations in Rust typically use Futures that map very well with JavaScript Promises. Rust is also an embeddable language, and that provides direct embedding capabilities to Deno. This added to Rust being one of the first languages to create a compiler for *WebAssembly*, made the Deno team choose it for its core.

Inspiration from POSIX systems

POSIX systems were of great inspiration to Deno. In one of his talks, Dahl even states that Deno handles some of its tasks *"as an operating system"*.

The following table shows some of the standard terms from POSIX/Linux systems and how they map to Deno concepts:

Posix/Linux	Deno
Processes	Web workers
Syscalls	Ops
File descriptors (fds)	Resource IDs (rids)
Scheduler	Tokio
Userland: libc++, glib, boost	`https://deno.land/std/`
/proc/stat	`Deno.metrics()`
man pages	Deno types

Some of the concepts from the Linux world might be familiar to you. Let's take, for instance, processes. They represent an instance of a running program that might execute using one or multiple threads. Deno uses WebWorkers to do the same job inside the runtime.

In the second row, we have syscalls. If you aren't familiar with them, they are the way for programs to perform requests to the kernel. In Deno, these requests do not go directly to the kernel; instead, they go from the Rust core to the underlying operating system, but they work similarly. We'll have the opportunity to see this in the upcoming architecture diagram.

These are just a couple of examples you might recognize if you are familiar with Linux/POSIX systems.

We'll explain and use most of the aforementioned Deno concepts throughout the rest of this book.

Architecture

Deno's core was initially written in *golang*, but it later changed to Rust. This decision was made to get away from *golang* as it is a garbage-collected language. Its combination with V8's garbage collector could lead to problems in the future.

To understand how the underlying technologies interact with each other to form the Deno core, let's look at the following architecture diagram for it:

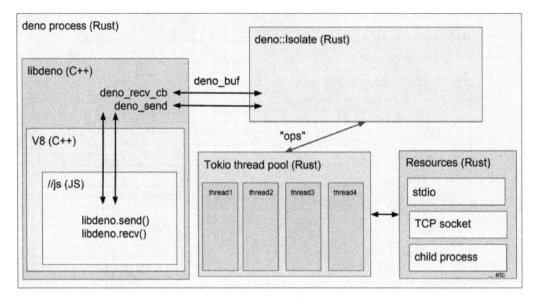

Figure 1.3 – Deno architecture

Deno uses message passing to communicate with the Rust backend. As a decision in regard to privilege isolation, Deno never exposes JavaScript object handles to Rust. All communication in and out of V8 uses Uint8Array instances.

For the event loop, Deno uses Tokio, a Rust thread pool. Tokio is responsible for handling I/O work and calling back the Rust backend, making it possible to handle all operations asynchronously. **Operations** (**ops**) is the name given to the messages that are passed back and forth between Rust and the event loop.

All the asynchronous messages dispatched from Deno's code into its core (written in Rust) return **Promises** back to Deno. To be more precise, asynchronous operations in Rust usually return **Futures**, which Deno maps to JavaScript Promises. Whenever these **Futures** are resolved, the JavaScript **Promises** are also resolved.

To enable communication from V8 to the Rust backend, Deno uses rusty_v8, a Rust crate created by the Deno team that provides V8 bindings to Rust.

Deno also includes the TypeScript compiler right inside V8. It uses V8 snapshots for startup time optimization. Snapshots are used for saving the JavaScript heap at a specific execution time and restoring it when needed.

Since it was first presented, Deno was subject to an iterative, evolutionary process. If you are curious about how much it changed, you can look at one of the initial roadmap documents written back in 2018 by Ryan Dahl (`https://github.com/ry/deno/blob/a836c493f30323e7b40e988140ed2603f0e3d10f/Roadmap.md`).

Now, not only do we know what Deno is, but we also know what's happening behind the scenes. This knowledge will help us in the future when we're running and debugging our applications. The creators of Deno made many technological and architectural decisions to bring Deno to the state it is today. These decisions pushed the runtime forward and made sure Deno excels in several situations, some of which we'll later explore. However, to make it work well for some use cases, some trade-offs had to be made. Those trade-offs resulted in the limitations we'll examine next.

Grasping Deno's limitations

As with everything, choosing solutions is a matter of dealing with trade-offs. The ones that adapt best to the projects and applications we're writing are what we end up using. Currently, Deno has some limitations; some of them due to its short lifetime, others because of design decisions. As it happens with most solutions, Deno is also not a one-size-fits-all tool. In the next few pages, we'll explore some of the current limitations of Deno and the motivations behind them.

Not as stable as Node.js

In its current state, Deno can't be compared to Node.js regarding stability for obvious reasons. Node.js has more than 10 years of development, while Deno is only nearing its second year.

Even though most of the core features presented in this book are already considered stable and correctly versioned, there are still features that are subject to change, and live under the unstable flag.

Node.js's years of experience made sure it is battle-tested and that it works in the most diversified environments. That's something we're hopeful Deno will get, but time and adoption are essential factors.

Better HTTP latency but worse throughput

Deno keeps performance on track from the beginning. However, as seen on the benchmarks page (`https://deno.land/benchmarks`), there are topics where it is still not at Node.js' level.

Its ancestor leverages the direct bindings with C++ on the HTTP server to amplify this performance score. Since Deno resisted to add native HTTP bindings and builds on top of native TCP sockets, it still suffers from a performance penalty. This decision is something that the team plans to tackle after optimizing TCP socket communication.

The Deno HTTP server handles about 25k requests per second with a max latency of 1.3 milliseconds, while Node.js handles 34k requests but has a latency that varies between 2 and 300 milliseconds.

We can't say 25k requests per second is not enough, especially since we're using JavaScript. If your app/website needs more than that, probably JavaScript, and thus Deno, is not the correct tool for the job.

Compatibility with Node.js

Due to many of the changes that have been introduced, Deno doesn't provide compatibility with existing JavaScript packages and tooling. A compatibility layer is being created on the standard library, but it is still not close to finished.

As Node.js and Deno are two very similar systems with shared goals, we expect the latter to execute more and more Node.js programs out of the box as time goes on. However, and even though some Node.js code is currently runnable, that is not the case currently.

TypeScript compiler speed

As we mentioned previously, Deno uses the TypeScript compiler. It reveals itself as one of the slowest parts of the runtime, especially compared to the time V8 takes to interpret JavaScript. Snapshots do help with this, but this is not enough. Deno's core team believes that they will have to migrate the TypeScript compiler to Rust to fix it.

Due to the extensive work required to complete this task, this is probably not going to happen anytime soon, even though it's supposed to be one of the things that would make its startup time orders of magnitude faster.

Lack of plugins/extensions

Even though Deno has a plugin system to support custom operations, it is not finished yet and is considered unstable. This means that extending native functionality to more than what Deno makes available is virtually impossible.

At this point, we should understand Deno's current limitations and why they exist. Some of them might be resolved soon, as Deno matures and evolves. Others are the result of design decisions or roadmap priorities. Understanding these limitations is fundamental when it comes to deciding if you should use Deno in a project. In the next section, we will have a look at the use cases we believe Deno is the perfect fit for.

Exploring use cases

As you are probably aware by now, Deno by itself has a lot of use cases in common with Node.js. Most of the changes that were made were to ensure the runtime is safer and more straightforward, but as it leverages most of the same pieces of technology, shares the same engine, and many of the same goals, the use cases can't differ by much.

However, and even though the differences are not that big, there may be small nuances that will make one a slightly better fit than the other in specific situations. In this section, we will explore some use cases for Deno.

A flexible scripting language

Scripting is one of those features where interpreted languages always shine. JavaScript is perfect when we want to prototype something fast. This can be renaming files, migrating data, consuming something from an API, and so on. It just feels like the right tool for these use cases.

Deno looked at scripting with much consideration. The runtime itself makes it very easy for users to write scripts with it, thus providing many benefits for this use case, especially compared to Node.js. These benefits are being able to execute code with just a URL, not having to manage dependencies, and the ability to create an executable based on Deno.

On top of all of this, the fact that you can now import remote code while controlling which permissions it uses is a significant step in terms of trust and security.

Deno's **Read Eval Print Loop** (**REPL**) is a great place to do experimentation work. Adding to what we mentioned previously, the small size of the binary and the fact it includes all the needed tools is the cherry on top of the cake.

Safer desktop applications

Although the plugin system is not stable yet and the packages that allow developers to create desktop applications depend heavily on that, it is very promising.

During the last few years, we've seen the rise of desktop web applications. The rise of the Electron framework (`https://www.electronjs.org/`) enabled applications such as VS Code or Slack to be created. These are web pages running inside a WebView with access to native features that are part of many people's daily lives.

However, for users to install these applications, they must trust them blindly. Previously, we discussed security and how JavaScript code used to have access to all the systems where it ran. Deno is fundamentally different here since, due to its sandbox and all its security features, this is much safer, and the potential that's unlocked is enormous.

We'll be looking at lots of advances in using JavaScript to build desktop applications in Deno throughout this book.

A quick and complete environment to write tools

Deno's features position it as a very complete, simple, and fast environment to write tooling in. When we say tooling, this is not only tooling for JavaScript or TypeScript projects. As the single binary contains everything needed to develop an application, we can use Deno in ecosystems outside of the JavaScript world.

Its clarity, automatic documentation via TypeScript, ease of running, and the popularity of JavaScript make Deno the right cocktail for writing tools such as code generators, automation scripts, or any other developer tools.

Running on embedded devices

By using Rust and distributing the core as a Rust crate, Deno automatically enables usage in embedded devices, from IoT devices to wearables and ARM devices. Again, the fact that it is small and includes all the tools in the binary might be a great win.

The fact that the crate is made available standalone allows people to embed Deno in different places. For instance, when writing a database in Rust and wanting to add Map-Reduce logic, we can use JavaScript and Deno to do so.

Generating browser-compatible code

If you haven't had a look at Deno before, then this probably comes as a surprise. Aren't we talking about a server-side runtime? We are. But this same server-side runtime has been making efforts to keep the API's browser compatible. It provides features in its toolchain that enable code to be written in Deno and executed in the browser, as we'll explore in *Chapter 7, HTTPS, Extracting Configuration, and Deno in the Browser.*

All of this is taken care of by the Deno team, which keeps its APIs browser-compatible and generates browser code that opens a new set of possibilities yet to be discovered. Browser compatibility is something we will use later in this book, in *Chapter 7*, *HTTPS, Extracting Configuration, and Deno in the Browser* to build a Deno application by writing a complete application, client, and server inside Deno.

Full-fledged APIs

Deno, like Node.js,, puts lots of effort into dealing with HTTP servers. With a complete standard library providing great primitives for frameworks to write on top of, there is no doubt that APIs are among the strongest Deno use cases. TypeScript is a great addition here in terms of documentation, code generation, and static type checking, helping mature code bases scale.

We'll be focusing more on this specific use case throughout the rest of this book as we believe it to be one of the most important ones – one where Deno shines.

These are just a few examples of use cases where we believe Deno is a great fit. As with Node.js, we're also aware that there are many new uses to discover. We're excited to accompany this adventure and see what it still has to unveil.

Summary

In this chapter, we traveled back in time to 2009 to understand the creation of Node.js. After that, we realized why and when we should use the event-driven approach compared to a threaded model and the advantages it brings. We came to understand what evented, asynchronous code is and how JavaScript helped Node.js and Deno make the most out of the server's resources.

After that, we fast-forwarded through the Node.js' 10+ year story, its evolution, and how its adoption started. We observed how the runtime grew, together with its base language, JavaScript, while helping millions of businesses deliver great products to its clients.

Then, we took a modern look at Node.js, with today's eyes. What changed in the ecosystem and the language? What are some of the developers' pain points? We dived into these pain points and explored why it was difficult and slow to change Node.js to solve them.

As this chapter progressed, Deno's motivations became more and more evident. After looking at the past of JavaScript on the server, it made sense for something new to appear – something that would solve the pain experienced previously while keeping the things developers love.

Finally, we got to know Deno, which will be our friend for this book. We learned its vision, principles, and how it offers to solve certain problems. After having a sneak peek at the architecture and the components that made Deno possible, we couldn't finish without talking about some of the trade-offs and current limitations.

We concluded this chapter by listing use cases where Deno is an excellent fit. We will come back to these use cases later in this book, when we start coding. From this chapter on, our approach will be more concrete and practical, always moving toward code and examples you can run and explore.

Now that we understand what Deno is, we have all it takes to start using it. In the next chapter, we will set up the respective environment and write a Hello World application, among doing many other exciting things.

That's how exciting adventures start, right? Let's go!

2
The Toolchain

Now that we're familiar with evented languages, Node's history, and the reasons that led to Deno, we're in good shape to start writing some code.

In this chapter, the first thing we'll do is set up the environment and code editor. We'll proceed by writing our first Deno program and using the REPL to experiment with the runtime APIs. Then, we'll explore the module system and how Deno cache and module resolution works with practical examples. We'll understand versioning, and we'll also learn how to handle third-party dependencies. Then, we'll use the CLI to explore packages and their documentation, as well as how to install and reuse Deno scripts.

After running and installing a few scripts, we'll dive into permissions by learning how the permission system works and how we can secure the code we run.

On our journey of learning about the toolchain, we can't leave code formatting and linting out, so we'll also explore these topics in this chapter. We'll explore Deno's test suite by writing and running some simple tests, and we'll finish by presenting how Deno can bundle code into a self-sustainable binary or a single JavaScript file.

In this chapter, we'll cover the following topics:

- Setting up the environment
- Installing VS Code
- Hello World

- The module system and third-party dependencies
- Running and installing scripts
- Using the test command
- Using permissions
- Formatting and linting code
- Bundling code
- Compiling to a binary
- Using the upgrade command

Let's get started!

Technical requirements

All the code present in this chapter can be found at `https://github.com/PacktPublishing/Deno-Web-Development/tree/master/Chapter02`.

Setting up the environment

One of Deno's principles is to keep its single single executable as complete as possible. This decision, among others, dramatically facilitates the installation step. In this section, we'll install VS Code and the recommended plugins and learn how to install Deno on different systems.

Installing Deno

In the next few pages, we'll learn how to install Deno. To make sure everything written in this book runs smoothly, we'll be using version 1.7.5.

This is one of the rare parts of this book where things might differ, depending on your operating system. After the installation is complete, it doesn't make a difference how you installed Deno.

Let's get practical and install Deno on our machines. The following bullet points show you how to install the runtime on different operating systems:

- **Shell (Mac, Linux)**:

```
$ curl -fsSL https://deno.land/x/install/install.sh | sh
-s v1.7.5
```

- **PowerShell (Windows)**:

```
$v="1.7.5"; iwr https://deno.land/x/install/install.ps1
-useb | iex
```

Then, to make sure everything worked, let's get the current Deno version by running the following command:

```
$ deno --version
```

We should get the following output:

```
$ deno --version

deno 1.7.5 (release, x86_64-apple-darwin)
v8 9.0.123
typescript 4.1.4
```

Now that we have the correct versions of Deno installed, we can start writing and executing our programs. However, to make our experience smoother, we will install and configure our editor of choice.

Installing VS Code

VS Code is the editor we will be using throughout this book. This is mainly because it has an official Deno plugin. There are other editors that provide a pleasant experience with JavaScript and TypeScript, so feel free to use them.

These series of steps are not required for following the rest of this book flawlessly, so feel free to skip them. To install it, follow these steps:

1. Go to https://code.visualstudio.com/ and click the **Download** button.

2. Once the download is complete, install it on your system.

3. Once VS Code is installed, the last step is to install Deno's VS Code plugin.

4. In the **Plugins** section (the fifth icon on the left sidebar of VS Code), search for
    Deno and install the Deno plugin authored by Denoland, which is the official one:

Figure 2.1 – Plugin icon on VS Code's left bar

This is what Deno's VS Code plugin looks like:

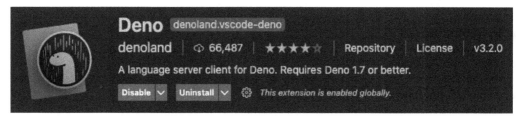

Figure 2.2 – Deno extension on the VS Code Marketplace

To enable the Deno plugin in your projects, you must create a local VS Code folder where
the workspace configuration files will live. To do that, we will create a folder named
.vscode with a file inside it called settings.json and write the following in that file:

```
{
  "deno.enable": true
}
```

This will make VS Code activate the extension inside the folder we're currently in. When
using unstable features, we can also enable the deno.unstable setting, which is also
mentioned in the plugin's documentation.

Shell completions

Deno also provides us with a way to generate shell completions. This way, we'll be getting
autocomplete suggestions when writing Deno commands on the Terminal. We can do this
by running the following command:

```
$ deno completions <shell>
```

The possible values for shell are `zsh`, `bash`, `fish`, `powershell`, and `elvish`. Make sure you choose the one you are using. This command will output the completions to stdout. You can then paste that into your shell profile (https://deno.land/manual@v1.7.5/getting_started/setup_your_environment#shell-autocomplete).

With that, we've finished how to install it Deno. We also have the runtime and the editor installed and configured. Now, let's write a Hello World program with Deno!

Hello World

With everything in place, let's write our first program!

First, we need to create a file named `my-first-deno-program.js` and write something that we're familiar with. We'll use the `console` API to write a message to the console:

```
console.log('Hello from deno');
```

To execute this, let's use the CLI we installed in the previous section. The command we must use to execute programs is called `run`:

```
$ deno run my-first-deno-program.js
Hello from deno
```

> **Tip**
> All Deno CLI commands can be executed with the `--help` flag, which will detail all the command's possible behaviors.

At this point, we haven't really done anything we don't know what to do already. We just wrote a `console.log` file in the language that we're familiar with, JavaScript.

The interesting thing is that we learned how to execute programs by using the `run` command. We'll explore this in more detail later in this chapter.

REPL

The **Read Eval Print Loop**, also known as the **REPL**, is a tool that is commonly used in interpreted languages. It allows users to run lines of code and get their immediate output. Node.js, Ruby, and Python are a few examples of languages where it is heavily used. Deno is no exception.

To open it, you just need to run the following command:

```
$ deno
```

You can now spend some time exploring the language (hint: there's tab completion). If you are curious about what APIs are available, you're in the right place to try them out. We'll deep dive into those later, but just to give you a few suggestions, you can look at the *Deno* namespace, Web API-compatible functions such as `fetch`, or objects such as `Math` or `window`, all of which are listed in Deno's documentation (`https://doc.deno.land/builtin/stable`).

Try them out!

Eval

Another way to execute code that doesn't live in a file is by using the `eval` command:

```
$ deno eval "console.log('Hello from eval')"
Hello from eval
```

The `eval` command can be useful to run simple, inline scripts.

So far, the programs we've have written have been quite simple. We just logged values to the console in a few different ways. However, as we start getting closer to the real world, we know we'll write more complex logic. With more complex logic comes more bugs and thus the need to debug our code. That's what we'll learn about next.

Debugging code in Deno

Even when we're following best practices and doing our best to write simple, clean code, any relevant program is very likely to need debugging once in a while.

Mastering the ability to quickly run and debug code is one of the best ways to improve your learning curve for any technology. This skill makes it easy to test and understand how stuff works by trial and error and fast experimentation.

Let's learn how can we debug our code.

The first step is to create a second program. Let's add a couple of variables that we can inspect later. The main objective of this program is to return the current time. We'll be using the already known `Date` object to do this. Let's call this file `get-current-time.js`, like so:

```
const now = new Date();
console.log(`${now.getHours()}:${now.getMinutes()}:
   ${now.getSeconds()}`);
```

What if we want to debug the value of the `now` variable before it prints to the console? This is where debugging comes in useful. Let's run the same program but with the `--inspect-brk` flag:

```
$ deno run --inspect-brk get-current-time.js
Debugger listening on ws://127.0.0.1:9229/ws/32e48d8a-5c9c-
4300-8e09-ee700ab79648
```

We can now open Google Chrome at `chrome://inspect/`. A remote target running on localhost called `deno` will be listed. By clicking on `inspect`, the Chrome DevTools inspector window will open and its execution will stop on the first line:

Figure 2.3 – Chrome stopping on the first line to be debugged

At this point, we can add breakpoints, log certain values, inspect variables, and so on. It enables the same as what we can do when we're debugging on Node or in the browser.

The `--inspect` flag could have also been used for this. However, we've used `--inspect-brk` here for convenience. Both have similar behaviors, but `inspect` requires a *debugger* to be present in the code. When the code executes and interprets the *debugger* keyword, it tries to connect to an already running inspector instance.

Now that we understand how to run and debug code, we can start writing our own programs. There's still much to learn, but we are already comfortable with the bare minimum.

As we start writing our programs and our code base grows, it is very likely that we'll start extracting logic into different modules. As these modules become reusable, we might extract them into packages so that they can be shared across projects. That's why we need to understand how Deno handles module resolution, which we'll do in the next section.

Modules and third-party dependencies

Deno uses ECMAScript modules and imports that are fully compatible with the browser. The path to a module is absolute, so it includes the file extension, which is also a standard in the *browser world*.

Deno takes the approach of being a *browser for scripts* quite seriously. One of the things it has in common with web browsers is that it deeply leverages URLs. They're one of the most flexible ways to share a resource and work beautifully on the web. Why not use them for module resolution? That's what browsers did.

The fact that the path for modules is absolute makes it possible not to depend on third-party entities such as npm, or complex module resolution strategies. With absolute imports, we can import code directly from GitHub, from a proprietary server, or even from a gist. The only requirement is that it has a URL.

This decision enables a completely decentralized module distribution to be used and makes module resolution inside Deno simple and browser compatible. This is something that doesn't happen on Node.

Deno even leverages URLs for versioning. For instance, to import version 0.83.0 the HTTP server from the standard library, we would use the following code:

```
import { serve } from
'https://deno.land/std@0.83.0/http/server.ts'
```

And that's how simple you can import a module. Here, the code is loading from `https://deno.land/`, but modules can be loaded from anywhere else. The only requirement is that there's a link to it.

For instance, if you have your own server where the files are available via a URL, you can directly use them within Deno. Previously, we learned that Deno automatically installs and caches dependencies, so let's learn a little more about how that works.

Locally cached dependencies

We already, we learned that Deno doesn't have conventions such as `node_modules`. For someone coming from Node, this might sound strange. Does this mean your code is always fetching modules from the internet? No. Can you still work offline? Yes.

Let's see this in practice.

Create a file called `hello-http-server.js` and add the following code there:

```
import { serve } from
"https://deno.land/std@0.84.0/http/server.ts";
for await (const req of serve(":8080")) {
   req.respond({ body: "Hello deno" });
}
```

As you can probably guess, this program starts an HTTP server at port `8080` and responds to every request made with `Hello deno`.

If that still feels strange to you, don't worry – we will have a more in-depth look at the standard library in the next chapter.

Let's run the program and pay attention to what Deno does before executing the code:

```
$ deno run hello-http-server.js
Download https://deno.land/std@0.83.0/http/server.ts
Download https://deno.land/std@0.83.0/encoding/utf8.ts
Download https://deno.land/std@0.83.0/io/bufio.ts
Download https://deno.land/std@0.83.0/_util/assert.ts
Download https://deno.land/std@0.83.0/async/mod.ts
Download https://deno.land/std@0.83.0/http/_io.ts
Download https://deno.land/std@0.83.0/textproto/mod.ts
Download https://deno.land/std@0.83.0/http/http_status.ts
Download https://deno.land/std@0.83.0/async/deferred.ts
```

```
Download https://deno.land/std@0.83.0/async/delay.ts
Download https://deno.land/std@0.83.0/async/mux_async_iterator.
ts
Download https://deno.land/std@0.83.0/async/pool.ts
Download https://deno.land/std@0.83.0/bytes/mod.ts
error: Uncaught PermissionDenied: network access to
"0.0.0.0:8080", run again with the --allow-net flag
```

What happened? Before it runs the code, Deno looks at the code's imports, downloads any dependencies, compiles them, and stores them in a local cache. There's still an error at the end, but we'll get to that later.

To understand what Deno does with the downloaded files, we'll use another command named `info`:

```
$ deno info
DENO_DIR location: "/Users/alexandre/Library/Caches/deno"
Remote modules cache: "/Users/alexandre/Library/Caches/deno/
deps"
TypeScript compiler cache: "/Users/alexandre/Library/Caches/
deno/gen"
```

This prints information about Deno's installation. Note DENO_DIR, which is the path where Deno is storing its local cache. If we navigate there, we can access the .js files and respective source maps.

After downloading and caching the modules for the first time, Deno will not redownload them and will keep using the local cache until it is explicitly asked not to.

Caching without running the code

To make sure you have a local copy of your code's dependencies without having to run it, you can use the following command:

```
$ deno cache hello-http-server.js
```

This will do the exact same thing that Deno does before running your code; the only difference is that it doesn't run. Due to this, we can establish some parallelism between the deno cache command and what npm install does on Node.

Reloading the cache

The cache and run commands can use the --reload flag to force the dependencies to be downloaded. A comma-separated list of modules that must be reloaded can be sent as a parameter to the --reload flag:

```
$ deno cache hello-http-server.js --reload=https://deno.land/
std@0.83.0/http/server.ts
Download https://deno.land/std@0.83.0/http/server.ts
```

In the preceding example, only modules from https://deno.land/std@0.83.0/http/server.ts would be redownloaded, as we can confirm by looking at the command's output.

Finally running the server

Now that the dependencies have been downloaded, we still have something preventing us from running the server, a PermissionDenied error:

```
error: Uncaught PermissionDenied: network access to
"0.0.0.0:8080", run again with the --allow-net flag
```

For now, let's follow the recommendation and add the --allow-net flag, which will grant our program full network access. We'll get into permissions later in this chapter:

```
$ deno run --allow-net hello-http-server.js
```

> **Tip (Windows)**
>
> Keep in mind that, if you are using Windows, you might get Windows' native network authorization popup, informing you that a program (Deno) is trying to access the network. If you want this web server to be able to run, you should click **Allow access**.

Now, our server should be running. If we curl to port 8080, it will show us Hello Deno:

```
$ curl localhost:8080
Hello deno
```

This wraps up for our simplest web server; we'll come back to this in a few pages.

Managing dependencies

If you have used other tools, or even Node.js itself, you might be feeling that it is not very intuitive to have URLs all around the code. We can also argue that, by directly writing URLs in our code, we might cause problems, such as having two different versions of the same dependency or having typos in the URL.

Deno solved this problem by getting rid of complex module resolution strategies and using plain JavaScript and absolute imports instead.

The proposed solution to keep track of dependencies, which is no more than a suggestion, is to use a file that exports all the required dependencies and place them in a single file that contains URLs. Let's see it in action.

Create a file called deps.js and add our dependencies there, exporting the ones we need:

```
export { serve } from
"https://deno.land/std@0.83.0/http/server.ts";
```

By using the preceding syntax, we're importing the serve method from the standard library's HTTP server and exporting it.

Back in our hello-http-server.js file, we can now change the imports so that we can use the exported function from the deps.js file:

```
import { serve } from "./deps.js";
for await (const req of serve(":8080")) {
  req.respond({ body: "Hello deno" });
}
```

Now, every time we add a dependency, we can run deno cache deps.js to guarantee that we have a local copy of the module.

This is Deno's way of managing dependencies. It is that simple – no magic, no complex standards, just a file that imports and exports symbols.

Integrity checking

Now that you know how to import and manage third-party dependencies, you might be feeling that something is still missing.

What can guarantee that the next time we, a coworker, or even a CI tries to install the project that our dependencies haven't changed?

That's a fair question, and since it's a URL, this might happen.

We'll can solve this by using integrity checking.

Generating a lock file

Deno has a feature that can store and check sub-resource integrity by using a JSON file, similar to the lock file approach that's used by other technologies.

To create our first lock file, let's run the following command:

```
$ deno cache --lock=lock.json --lock-write deps.js
```

Using the --lock flag, we select the name of the file, and by using --lock-write, we're giving Deno permission to create or update that same file.

Looking at the generated lock.json file, this is what we'll find there:

```
{
    "https://deno.land/std@0.83.0/_util/assert.ts":
    "e1f76e77c5ccb5a8e0dbbbe6cce3a56d2556c8cb5a9a8802fc9565
af72462149",
    "https://deno.land/std@0.83.0/async/deferred.ts":
    "ac95025f46580cf5197928ba90995d87f26e202c19ad961bc4e317
7310894cdc",
    "https://deno.land/std@0.83.0/async/delay.ts":
    "35957d585a6e3dd87706858fb1d6b551cb278271b03f52c5a2cb70
e65e00c26a",
```

It generates a JSON object where the keys are the paths to dependencies and the values are the hashes Deno uses to warrant resource integrity.

This file should then be checked into your version control system.

In the next section, we'll learn how to install dependencies and make sure everyone is running the exact same version of the code.

Installing dependencies with a lock file

Once the lock file has been created, anyone that wants to download the code can run the cache command with the --lock flag. This enables integrity checking while you're downloading dependencies:

```
$ deno cache --reload --lock=lock.json deps.js
```

It is also possible to use the `--lock` flag with the `run` command to enable runtime verification:

```
$ deno run --lock=lock.json --allow-net hello-http-server.js
```

> **Important note**
>
> When using the lock flags with the `run` command, code containing dependencies that haven't been cached yet will not be checked against the lock file.

To make sure the new dependencies are checked at runtime, we can use the `--cached-only` flag.

This way, Deno will throw an error if any dependencies that are not in the `lock.json` file are used by our code.

And that's all we need to do to make sure we're running the exact versions of the dependencies we want, eliminating any problems that might come out of version changes.

Import maps

Deno supports import maps (`https://github.com/WICG/import-maps`).

If you are not familiar with what they are, I'll explain them briefly for you: they are used to control JavaScript imports. If you've used JavaScript with code bundlers such as webpack before, then this is a feature similar to what you know as "aliases".

> **Important note**
>
> This feature is currently unstable, so it must be enabled using the `--unstable` flag.

Let's create a JSON file. The name doesn't matter here, but we'll name it `import-maps.json` for simplicity.

Inside this file, we'll create a JavaScript object with the `imports` key. In this object, any keys will be the module names and any values will be the real import paths. Our first *import map* will be one that maps the `http` word to the standard library HTTP module's root:

```
{
  "imports": {
    "http/": "https://deno.land/std@0.83.0/http/"
```

```
  }
}
```

By doing this, we can now import the standard library's HTTP module into our `deps.js` file, like so:

```
export { serve } from "http/server.ts";
```

To run it, we will use the `--import-map` flag. By doing this, we can select the file where the import maps are. Then, because this feature is still unstable, we must use the `--unstable` flag:

```
$ deno run --allow-net --import-map=import-maps.json --unstable
hello-http-server.js
```

As we can see, our code runs perfectly.

This is an easy way to customize module resolution that doesn't depend on any external tools. It's also been proposed as something to be added to browsers. Hopefully, this will be accepted in the near future.

Inspecting modules

We just used the standard library's HTTP module to create a server. Don't worry if you are not very familiar with the standard library yet; we'll explain it in more detail in the next chapter. For now, all we need to know is that we can explore its modules on its website (`https://deno.land/std`).

Let's take a look at the module we used in the previous script, the HTTP module, and use Deno to find out more information about it.

We can use the `info` command to do this:

```
$ deno info https://deno.land/std@0.83.0/http/server.ts
local:
/Users/alexandre/Library/Caches/deno/deps/https/deno.
land/2d926cfeece184c4e5686c4a94b44c9d9a3ee01c98bdb4b5e546dea4
e0b25e49
type: TypeScript
compiled: /Users/alexandre/Library/Caches/deno/gen/https/deno.
land/2d926cfeece184c4e5686c4a94b44c9d9a3ee01c98bdb4b5e546dea4
e0b25e49.js
deps: 12 unique (total 63.31KB)
```

```
https://deno.land/std@0.83.0/http/server.ts (10.23KB)
├── https://deno.land/std@0.83.0/_util/assert.ts *
├─┬ https://deno.land/std@0.83.0/async/mod.ts (202B)
│ ├── https://deno.land/std@0.83.0/async/deferred.ts *
│ ├── https://deno.land/std@0.83.0/async/delay.ts (279B)
│ ├─┬
...
│     └── https://deno.land/std@0.83.0/encoding/utf8.ts *
└─┬ https://deno.land/std@0.83.0/io/bufio.ts (21.15KB)
    https://deno.land/std@0.83.0/_util/assert.ts (405B)
    https://deno.land/std@0.83.0/bytes/mod.ts (4.34KB)
```

This command lists a great amount of information about the HTTP module. Let's break it down.

On the first line, we get the path to the cached version of the script. On the line after that, we see the type of the file. We already know that the standard library was written in TypeScript, so this should be of no surprise to us. The next line is also a path, this time for the compiled version of the module, since TypeScript modules are compiled to JavaScript in the download step.

The last part of the command's output is the dependency tree. By looking at it, we can quickly identify that it just links to other modules in the standard library.

Tip

We can use the --unstable and --json flags together with deno info to get a programmatically accessible JSON output.

When using third-party modules, more than knowing what they depend on, we need to know what functions and objects are made available by the module. We'll learn about this in the next section.

Exploring the documentation

Documentation is a crucial aspect of any software project. Deno does a good job of keeping all the APIs well-documented, and TypeScript helps a lot with this. As the standard library and runtime functions are all written in TypeScript, most of the documentation is automatically generated.

The documentation is available at https://doc.deno.land/.

If you don't have internet access and want to access the documentation of a module you have installed locally, Deno has got you covered.

Many editors, namely VS Code, allow you to do this, with the famous *Cmd/Ctrl* + click being an example. However, Deno doesn't depend on editor features for this as the doc command provides all the essential features you'll need.

Let's have a look at the documentation for the standard library's HTTP module:

```
$ deno doc https://deno.land/std@0.83.0/http/server.ts
function _parseAddrFromStr(addr: string): HTTPOptions
    Parse addr from string


async function listenAndServe(addr: string | HTTPOptions,
handler: (req: ServerRequest) => void): Promise<void>
    Start an HTTP server with given options and request handler


async function listenAndServeTLS(options: HTTPSOptions,
handler: (req: ServerRequest) => void): Promise<void>
    Start an HTTPS server with given options and request
      handler


function serve(addr: string | HTTPOptions): Server
    Create a HTTP server


...
```

We can now see the exposed methods and types.

In one of our previous programs, we used the serve method. To find out more about this specific method, we can send the method (or any other symbol) name as a second argument:

```
$ deno doc https://deno.land/std@0.83.0/http/server.ts serve
Defined in https://deno.land/std@0.83.0/http/server.ts:282:0


function serve(addr: string | HTTPOptions): Server
    Create a HTTP server


        import { serve } from
```

```
    "https://deno.land/std/http/server.ts";
    const body = "Hello World\n";
    const server = serve({ port: 8000 });
    for await (const req of server) {
      req.respond({ body }); add
    }
```

This is a very useful feature that enables developers to navigate the documentation of the locally installed modules without having to depend on an editor.

As we will learn in the next chapter, and as you probably noticed by using the REPL, Deno has a built-in API. To check out its documentation, we can run the following command:

```
$ deno doc --builtin
```

The output will be massive as it lists all the public methods and types.

In *nix systems, this can easily be piped into an application such as less:

```
$ deno doc --builtin | less
```

Similar to remote modules, it's also possible to filter by method name. Take, for instance, the writeFile function that's present in Deno's namespace:

```
$ deno doc --builtin Deno.writeFile
Defined in lib.deno.d.ts:1558:2

function writeFile(path: string | URL, data: Uint8Array,
options?: WriteFileOptions): Promise<void>
  Write `data` to the given `path`, by default creating a new
file if needed,
  else overwriting.

  ```ts
  const encoder = new TextEncoder();
  const data = encoder.encode("Hello world\n");
  await Deno.writeFile("hello1.txt", data);  // overwrite
"hello1.txt" or create it
  await Deno.writeFile("hello2.txt", data, {create: false});
// only works if "hello2.txt" exists
  await Deno.writeFile("hello3.txt", data, {mode: 0o777});  //
```
```

```
set permissions on new file
  await Deno.writeFile("hello4.txt", data, {append: true});  //
add data to the end of the file
  ```
```

```
Requires `allow-write` permission, and `allow-read` if
`options.create` is `false`.
```

The doc command is a useful part of the development workflow. However, if you have internet access and want to access it in a more digestible and visual way, https://doc. deno.land/ is the place to go.

You can use the documentation website to find out more about the built-in APIs or standard library modules. In addition, it also allows you to display the documentation for any module that's available. To do this, we just need to replace the :// part of the module's URL with a single backslash, /, and prepend https://doc.deno.land/ to the URL.

For instance, to access the documentation of the HTTP module, the URL would be https://doc.deno.land/https/deno.land/std@0.83.0/http/server.ts.

If you navigate to that URL, a clean interface will be displayed containing the module's documentation.

We now know how to use and explore third-party modules. However, as we start writing our applications, there might be some utilities that we want to share across projects. We may also want to have that specific package available everywhere in our system. The next section will help us do that.

Running and installing scripts

In one of his first talks, and in Deno' first version release notes (https://deno.land/ posts/v1#a-web-browser-for-command-line-scripts) Dahl used a sentence I like a lot:

"Deno is like a web browser for command-line scripts."

Every time I use Deno, this sentence makes more and more sense to me. I'm sure it will also start to make sense for you as the book proceeds. Let's explore it a little further.

In a browser, when you access a URL, it runs the code that is there. It interprets the HTML and the CSS, and then executes some JavaScript.

Deno, by following its premise of being a browser for scripts, just needs a URL to run code. Let's see it in action.
```

Honestly, it is not very different from what we've already done a couple of times already. As a refresher, the last time we executed our simple web server, we did the following:

```
$ deno run --allow-net --import-map=import-maps.json --unstable
hello-http-server.js
```

Here, `hello-http-server.js` was just a file in the current folder.

Let's try to do the same but with a remote file – a file that is served through HTTP.

We'll execute an "echo server" from the Deno standard library's set of examples. You can check out the code for this here (`https://deno.land/std@0.83.0/examples/echo_server.ts`). It is a server that echoes back whatever is sent to it:

```
$ deno run --allow-net https://deno.land/std@0.83.0/examples/
echo_server.ts
Download https://deno.land/std@0.83.0/examples/echo_server.ts
Check https://deno.land/std@0.83.0/examples/echo_server.ts
Listening on 0.0.0.0:8080
```

> **Important note**
>
> If you're using a Windows machine, it might not be possible to access `0.0.0.0:8080`; you should access `localhost:8080` instead. They both refer to the same thing in your local machine. However, when `0.0.0.0` appears throughout the rest of book, you should try to access `localhost` if you're running Windows.

As it happens, every time files are not cached, Deno downloads and executes them.

Does it differ that much from a web browser? I wouldn't say so. We gave it a URL, and it ran the code.

To make sure it is working, we can establish a Telnet (`https://en.wikipedia.org/wiki/Telnet`) connection and send a message that the server echoes back:

```
$ telnet 0.0.0.0 8080
Trying 0.0.0.0...
Connected to 0.0.0.0.
Escape character is '^]'.
hello buddy
hello buddy
```

You can do this with any available Telnet client; here, we're using a macOS client that we installed via Homebrew (`https://brew.sh/`). The first "hello buddy" is the message we sent, while the latter is the one that was echoed back. With this, we can verify that the echo server is working.

> **Important note**
>
> If you are using any other telnet client, make sure you enable "Local line editing" setting. Some clients don't have this enabled by default and send characters as you type them, resulting in a message with duplicated characters. The image below shows you how can you configure that setting in PuTTY for Windows.

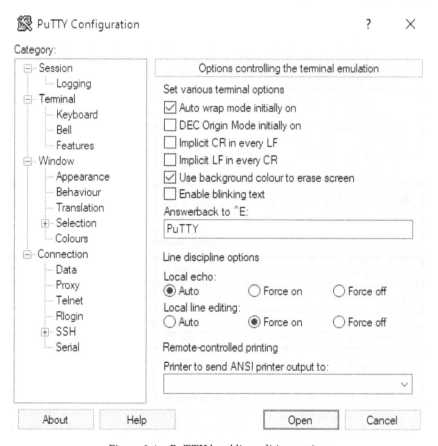

Figure 2.4 – PuTTY local line editing setting

This confirms what we previously stated, in that Deno uses the same approach to run code it uses to resolve modules: it treats local and remote codes in a similar way.

Installing utility scripts

There are utility programs we write once and those that we use multiple times. Sometimes, to facilitate reusing them, we just copy those scripts from project to project. For others, we keep them in a GitHub repository and keep going there to get them. The ones we use the most might need to be wrapped in a shell script, added it /usr/local/bin (on *nix systems) and made usable across our system.

For this, Deno provides the install command.

This command wraps a program into a thin shell script and puts it into the installation bin directory. The permissions for the script are set at the time of its installation and never asked for again:

```
$ deno install --allow-net --allow-read https://deno.land/
std@0.83.0/http/file_server.ts
```

Here, we used another module from the standard library called file_server. It creates an HTTP server that serves the current directory. You can see its code by accessing the import URL (https://deno.land/std@0.83.0/http/file_server.ts).

The installation command will make the file_server script available on your system.

To give it a name other than file_server, we can use the -n flag, like so:

```
$ deno install --allow-net --allow-read -n serve https://deno.
land/std@0.83.0/http/file_server.ts
```

Now, let's serve the current directory:

```
$ serve
HTTP server listening on http://0.0.0.0:4507/
```

If we access http://localhost:4507, this is what we'll get:

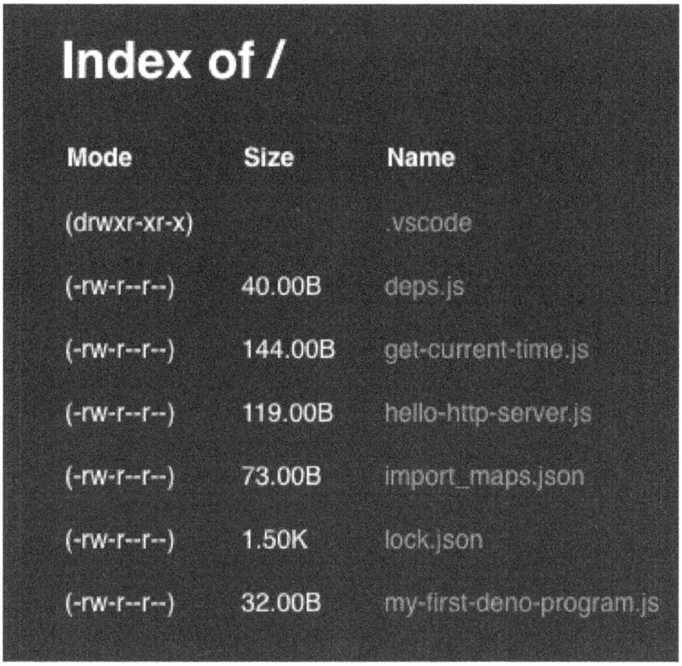

Figure 2.5 – Deno file server web page

This works with remote URLs but can also work with local ones. If you have a program written in Deno that you want to transform into an executable, you can also use the `install` command to do so.

We can do this with our simple web server, for instance:

```
$ deno install --allow-net --unstable hello-http-server.js
```

By running the preceding code, a script called `hello-http-server` is created and available across our system.

That's all we need to run in order to execute local and remote scripts. Deno makes this very easy because it treats imports and modules in a very straightforward way, very much like browsers do.

Previously, we've used permissions to allow scripts to access resources like network or the filesystem. In this section, we've used permissions with the `install` command, but we previously did this with the `run` command too.

By now, you probably understand how they work, but we'll look at them in more detail in the following section.

Permissions

We came across Deno's permissions for the first time a few pages ago when we wrote our first HTTP server. At the time, we had to give our script permission to access the network. Since then, we've used them a few times without knowing too much about how they work.

In this section, we'll explore how permissions work. We'll learn what permissions exist and how to configure them.

If we run `deno run --help`, we get the help output for the `run` command, which is where, among other things, certain permissions are listed. To make this easier for you, we will list all the existing permissions and provide a brief explanation of each.

-A, --allow-all

This disables all permission checks. Running code with this flag means it will have access to everything the user has, quite similar to what happens with Node.js by default.

Be careful when you run code with this, and be especially careful when the code is not yours.

--allow-env

This grants access to the environment. It's used so that programs can access environment variables.

--allow-hrtime

This grants access to high-resolution time management. It can be used for precise benchmarking. Giving this permission to the wrong scripts can allow for fingerprinting and timing attacks.

--allow-net=<domains>

This grants access to the network. When used without arguments, it allows all network access. When used with arguments, it allows us to pass a comma-separated list of domains where network communication will be allowed.

--allow-plugin

This allows plugins to be loaded. Note that this is still an unstable feature.

--allow-read=<paths>

This grants read access to the filesystem. When used without arguments, it grants access to everything the user has access to. With arguments, this only allows access to the folders provided by a comma-separated list.

--allow-run

This grants access to running subprocesses (for instance, with `Deno.run`). Keep in mind that subprocesses are not sandboxed and should be used with caution.

--allow-write=<paths>

This grants filesystem write access. When used without arguments, it grants access to everything the user has access to. With arguments, it only allows access to folders provided by
a comma-separated list.

Every time a program runs and it doesn't have the correct permissions,
a `PermissionError` will be thrown.

Permissions are used on the `run` and `install` commands. The only difference between them is the moment you give the permissions. For `run`, you have to give them when running, while for `install`, you give them when you install the script.

There is another way for a Deno program to have permissions. It doesn't require the permissions to be given upfront, and instead will ask for them as they're needed. We'll explore this feature in the next chapter, where we'll learn about Deno's namespace.

And that's it! There's really not much to add on permissions, except that they're a very important feature in Deno as it sandboxes our code by default and lets us decide on what should our code have access to. We'll keep using permissions on the applications we'll be writing throughout this book.

So far, we've learned how to run, install, and cache modules, as well as how to use permissions. As we write and run more complex programs, the need to test them starts to arise. We can do this with the `test` command, as we'll learn in the next section.

Using the test command

Included as part of the main binary, Deno also provides a test runner. The command for it, not surprisingly, is called `test`. In this section, we'll explore it and run a couple of tests.

In this section, we'll mainly explore the command itself and not the test syntax. We'll look at the syntax and best practices for it in more depth in a dedicated chapter later on the book.

The `test` command finds files to run based on the `{*_,*.,}test.{js,mjs,ts,jsx,tsx}` glob expression.

Since glob expressions might not be too intuitive to read, we'll explain them briefly.

It matches any files with the `js`, `mjs`, `ts`, `jsx`, and `tsx` extensions and that have `test` in their name preceded by an underscore (`_`) or a dot (`.`)

A few examples of files that will match the expression and be considered tests are as follows:

- `example.test.ts`
- `example_test.js`
- `example.test.jsx`
- `example_test.mjs`

Deno tests also run inside the sandbox environment, so they need permissions. Take a look at the previous section to find out more about how to do this.

When running tests, it is also possible to use the debugging commands we learned about earlier in this chapter.

Filtering tests

A common need when you have a complete test suite is to run only a specific part of it. For this, the test command provides the --filter flag.

Imagine that we have the following file, with two tests defined:

```
Deno.test("first test", () => {});
Deno.test("second test", () => {});
```

If we want to run just one of them, we can use the --filter flag and pass a string or pattern that will match the test names:

```
$ deno test --filter second
running 1 tests
test second test ... ok (3ms)

test result: ok. 1 passed; 0 failed; 0 ignored; 0 measured; 1
filtered out (3ms)
```

The preceding code just ran the test that matched the filter. This feature becomes very useful when we're developing tests for a small part of the code base and we want to get quick feedback about the process.

Fail fast

In environments such as a continuous integration server, we might want to fail fast if it doesn't really matter how many tests are failing and we just want the testing phase to be over if any test fails.

To do this, we can use the --fail-fast flag.

And that's all we need to know about testing for now. As we mentioned previously, we'll get back into testing in *Chapter 8, Testing - Unit and Integration*. We just wanted to become familiar with the CLI command here.

We consider tests as a tool that guarantees that our code is working but also as a way to document our code's behavior. Tests are fundamental in any working and evolving code base, and Deno makes them first-class citizens by including a test runner in its binary. However, testing is just a part of a bigger toolset – one where developer needs such as linting and formatting are also covered.

In the next section, we'll learn how Deno solves these problems.

Formatting and linting

Linting and formatting are two capacities considered crucial for maintaining consistency and enforcing good practices in any code base. With this in mind, Deno has embedded the tools to enable both in its CLI. We'll get to know them in this section.

Formatting

To format Deno's code, the CLI provides the `fmt` command. This is an opinionated formatter that aims to solve any questions regarding code formatting. The main goal is for developers to not have to care about the format of their code – not when writing code nor when reviewing pull requests.

Running the following command with no argument formats all the files in the current directory:

```
$ deno fmt
/Users/alexandre/Deno-Web-Development/Chapter02/my-first-deno-
program.js
/Users/alexandre/Deno-Web-Development/Chapter02/bundle.js
```

If we want to format a single file, we can send it as an argument.

To check files for formatting errors, we can use this together with the `--check` flag, which will output the errors that were found in our files to stdout.

Ignoring lines and files

To make the formatter ignore a single line or a complete file, we can use the `ignore` comment:

```
// deno-fmt-ignore
const book = 'Deno 1.x - Web Development';
```

Using deno-fmt-ignore ignores the line right after the comment:

```
// deno-fmt-ignore-file
const book = 'Deno 1.x - Web Development';


const editor = 'PacktPub'
```

Using deno-fmt-ignore-file ignores the whole file.

Lint

Still under the unstable flag, the lint command prints warnings and errors found in our code to stdout.

Let's see it in action by running the linter against a script called to-lint.js. You can run it against whatever you want. Here, we just a file that will throw an error since it contains a debugger:

```
$ deno lint --unstable to-lint.js
(no-debugger) `debugger` statement is not allowed
   debugger;
   ~~~~~~~~~
      at /Users/alexandre/dev/personal/Deno-Web-Development/
Chapter02/to-lint.js:4:2
Found 1 problems
```

In this section, we learned how do use the fmt and lint commands to enforce code consistency and best practices across a code base.

These are two of the commands that are provided by the Deno CLI that will be used in our daily life when writing Deno programs. They both happen to be very opinionated, so there is no space to support different standards. This should be of no surprise since Deno is heavily inspired by *golang*, and this approach is in line with what can be done by tools such as *gofmt*.

With that, we know how to format and check our code for best practices. Adding this to what we learned in the previous sections, there is nothing stopping us from running our code in production.

When we get into production, we obviously want our servers to be as fast as possible. In the previous chapter, we learned that one of slowest parts of Deno is TypeScript parsing. When we write TypeScript code, we don't want to sacrifice time parsing it every time a new instance of the server starts up. At the same time, as we write clean, separate modules, we don't want to ship them separately into the production environment.

This is why Deno provides a feature that allows us to bundle code into a single file. We'll learn about this in the next section.

Bundling code

In the previous chapter, when we presented Deno, we elected bundling code as an exciting feature for many reasons. This feature has enormous potential, and we will explore this in more detail in *Chapter 7, HTTPS, Extracting Configuration, and Deno in the Browser*. But since we're exploring the CLI here, we'll get to know the appropriate command.

It is called bundle, and it bundles code into a single, self-contained ES module.

Bundled code that doesn't depend on the Deno namespace can also run in the browser with <script type="module"> and in Node.js.

Let's use it to build our get-current-time.js script:

```
$ deno bundle get-current-time.js bundle.js
Bundle file:///Users/alexandre/dev/deno-web-development/
Chapter02/2-hello-world/get-current-time.js
Emit "bundle.js" (2.33 KB)
```

Now, we can run the generated bundle.js:

```
$ deno run bundle.js
0:11:4
```

This will print the current time.

We can also execute it with Node.js since it is ES6 compatible JavaScript (you need to have Node.js installed to be able to run the following command):

```
$ node bundle.js
0:11:4
```

To use this same code in the browser, we can create a file named `index-bundle.html` and import our generated bundle:

```html
<!DOCTYPE html>
<html>
  <head>
    <title>Deno bundle</title>
  </head>
  <body>
    <script type="module" src="bundle.js"></script>
  </body>
</html>
```

With the knowledge we gained from the previous section, we can run the standard library's file server in the current folder:

```
$ deno run --allow-net --allow-read https://deno.land/
std@0.83.0/http/file_server.ts
HTTP server listening on http://0.0.0.0:4507/
```

Now, if you navigate to `http://localhost:4507/index-bundle.html`, and open the browser console, you'll see that the current time has been printed..

Bundling is a very promising feature that we will explore later in *Chapter 7, HTTPS, Extracting Configuration, and Deno in the Browser*. It allows us to create a single JavaScript file with our application.

We'll get back to this and show you what things it enables later on in this book. Bundling is a nice way to distribute your Deno applications, as we saw in this chapter. But what if you want to distribute your application so that it can run in places that are not your computer? Does the `bundle` command do this for us?

Well, not really. It does this if the place where the code will be executed has Node, Deno, or a browser installed.

But what if it doesn't? That's what we'll learn about next.

Compiling to a binary

When Deno was initially launched, Dahl stated that one of its objectives was to be able to ship Deno code as a single binary, something similar to what golang does, from day one. This is very similar from the work that nexe (`https://github.com/nexe/nexe`) or pkg (`https://github.com/vercel/pkg`) do for Node.

This is different from the bundle feature, where a JavaScript file is generated. What happens when you compile Deno code to a binary is that all the runtime and code is included in that binary, making it self-sustainable. Once you've compiled it, you can just send this binary anywhere, and you'll be able to execute it.

> **Important note**
>
> At the time of writing, this is still an unstable feature with many limitations, as mentioned in its release notes at `https://deno.land/posts/v1.7#improvements-to-codedeno-compilecode`.

This process is very simple. Let's see how we can do this.

We just need to use the `compile` command. For this example, we'll use the script we used in the previous section; that is, `get-current-time.js`:

```
$ deno compile --unstable get-current-time.js
Bundle file:///Users/alexandre/dev/Deno-Web-Development/
Chapter02/get-current-time.js
Compile file:///Users/alexandre/dev/Deno-Web-Development/
Chapter02/get-current-time.js
Emit get-current-time
```

This emits a binary named `get-current-time`, which we can now execute:

```
$ ./get-current-time
16:10:8
```

It's working! This feature allows us to easily distribute the application. This is possible because it includes the code and all its dependencies, including the Deno runtime, making it self-sustainable.

As Deno continues to evolve, new features, bug fixes, and improvements will be added. At its current pace of a few releases per quarter, it's very common that you might want to upgrade our version of Deno. The CLI also provides a command for this. We'll learn about this in the next section.

Using the upgrade command

We started this chapter by learning how to install Deno, and we installed a single version of the runtime. But Deno is constantly shipping bug fixes and improvements – even more so in these early versions.

When there are new updates, we can use the same package manager we used to install Deno to upgrade it. However, the Deno CLI provides a command it can use to upgrade itself. The command is called upgrade and can be used together with the --version flag to select the version we're upgrading to:

```
$ deno upgrade --version=1.7.4
```

If no version is provided, it defaults to the latest version. To install the newer version in another location, instead of replacing the current installation, you can use the --output flag, like so:

```
$ deno upgrade --output $HOME/my_deno
```

And that's it – upgrade is one more utility that follows the Deno philosophy of providing all we need to write and maintain applications, and part of that cycle is definitely updating our runtime.

Summary

In this chapter, our main focus was to get to know the tools Deno provides, including those in its main binary. These tools will be heavily used in our daily life and throughout the rest of this book.

We started by getting our environment and editor in place and then deep dived into the toolchain.

Then, we wrote and executed a **Hello World** application. The **REPL** and the eval command were presented as ways to enable experimentation and running code without a file. After that, we look at the module system. We not only imported and used modules, but we also looked under the hood and understood how Deno downloads and caches dependencies locally.

After become familiar with the module system, we learned about how to manage external dependencies, namely lock files and integrity checking. We couldn't leave this section without speaking a little about a still unstable but promising feature: import maps.

After that, we explored some third-party modules and their code and dependencies with the help of the `info` command. The documentation wasn't left on the side by Deno, and we also learned how we can use the `documentation` command and the respective website to look at third-party's code documentation.

Since scripting is a first-class citizen in Deno, we explored the commands that enable us to reuse code by directly running code from a URL and installing utility scripts globally.

Throughout this book, we've mentioned that permissions are one of Deno's highlights. In this chapter, we learned how to use permissions when running code to fine-tune its privileges.

Next, we learned about the test runner, and how to run and filter tests. Another feature we learned about was how to format and lint our code according to Deno's standards. We got to know the `fmt` and `lint` commands, two opinionated tools that ensure that developers don't have to be concerned about formatting and linting since they're handled automatically.

Finally, we introduced the `bundle` and `compile` commands. We learned how we can bundle our code into a single JavaScript file, and how to generate a binary that includes our code and the Deno runtime, making it self-sustainable.

A lot of interesting ground was covered in this chapter. I promise you that what's next is even more exciting. In the next chapter, we'll get to know the standard library and write simple applications with it while we learn about Deno's APIs.

Excited? Let's go!

3
The Runtime and Standard Library

Now that we know enough about Deno, we're in a good place to write a few real applications with it. In this chapter, we'll be using no libraries as its primary purpose is to present the runtime APIs and the standard library.

We will be writing small CLI utilities, web servers, and more, always leveraging the power of what the official Deno team created, with no external dependencies.

The Deno namespace will be our starting point as we believe it makes sense to explore what the runtime includes first. Following this idea, we'll also look at the Web APIs that Deno shares with the browser. We'll use `setTimeout` to `addEventListener`, `fetch`, and so on.

Still in the Deno namespace, we will get to know the program lifecycle, interact with the filesystem, and build small command-line programs. Later, we will get to know buffers and understand how they can be used to asynchronously read and write.

We will then take a quick turn into the standard library and we'll go through some useful modules. This chapter doesn't aim to replace the standard library's documentation; it will instead present you with some of its capabilities and use cases. We'll get to know it while we write small programs.

On this journey through the standard library, we will use modules that deal with the filesystem, ID generation, text formatting, and HTTP communication. Part of it will be an introduction to what we'll explore in more depth in later chapters. You'll finish this chapter by writing your first JSON API, and connecting to it.

The following are the topics that we will be covering in this chapter:

- The Deno runtime
- Exploring the Deno namespace
- Using the standard library
- Building a web server using the HTTP module

Technical requirements

All the code files of this chapter can be found at the following GitHub link: `https://github.com/PacktPublishing/Deno-Web-Development/tree/master/Chapter03`.

The Deno runtime

Deno provides a set of functions that are included in the runtime as globals in the `Deno` namespace. The runtime APIs are documented at `https://doc.deno.land/` and can be used to do the most elementary, low-level things.

Two types of functions are available on Deno without any imports: Web APIs and the `Deno` namespace. Whenever there's a behavior in Deno that also exists on the browser, Deno mimics the browser APIs – those are Web APIs. Since you come from the JavaScript world, you're probably familiar with most of them. We're speaking about functions such as `fetch`, `addEventListener`, `setTimeout`, and objects such as `window`, `Event`, `console`, among others.

Code written using Web APIs can be bundled and run in the browser with no transformations.

The other big part of the APIs exposed by the runtime lives inside a global namespace named Deno. You can use the REPL and the documentation, two of the things we explored in *Chapter 2, The Toolchain*, to explore it and get a quick grasp of what functions it includes. Later in this chapter, we'll also experiment with some of the most common ones.

If you want to access the documentation of all the symbols that are included in Deno, you can run the doc command with the --builtin flag.

Stability

The functions inside the Deno namespace are considered stable from version 1.0.0 onwards. This means the Deno team will make an effort to support them across newer versions, and will do its best to keep them compatible with future changes.

Features that are still not considered stable for production live under the --unstable flag, as you probably imagined, since we've used them in previous examples.

The documentation of unstable modules can be accessed by using the --unstable flag with the doc command or by accessing https://doc.deno.land/builtin/unstable.

The standard library is not yet considered stable by the Deno team and thus they have a different version from the CLI (at the time of writing, it is on version 0.83.0).

In contrast with the Deno namespace functions, the standard library doesn't normally need the --unstable flag to run, except if any module from the standard library is using unstable functions from the Deno namespace.

Program lifecycle

Deno supports the browser compatible load and unload events that can be used to run setup and cleanup code.

Handlers can be written in two different ways: with addEventListener and by overriding the window.onload and window.onunload functions. The load events can be asynchronous but the same is not true for unload events as they can't be canceled.

Using addEventListener enables you to register unlimited handles; for instance:

```
addEventListener("load", () => {
  console.log("loaded 1");
});
addEventListener("unload", () => {
  console.log("unloaded 1");
```

```
});
addEventListener("load", () => {
  console.log("loaded 2");
});
addEventListener("unload", () => {
  console.log("unloaded 2");
});
console.log("Exiting...");
```

If we run the preceding code, we get the following output:

```
$ deno run program-lifecycle/add-event-listener.js
Exiting...
loaded 1
loaded 2
unloaded 1
unloaded 2
```

Another way to schedule code to run on setup and teardown phases is by overriding the `onload` and `onunload` functions from the `window` object. These functions have the particularity that only the last to be assigned runs. This happens because they override one another; see the following code, for instance:

```
window.onload = () => {
  console.log("onload 1");
};
window.onunload = () => {
  console.log("onunload 1");
};
window.onload = () => {
  console.log("onload 2");
};
window.onunload = () => {
  console.log("onunload 2");
};
console.log("Exiting");
```

By running the preceding program, we got the following output:

```
$ deno run program-lifecycle/window-on-load.js
Exiting
onload 2
onunload 2
```

If we then look at the initial code we wrote, we can understand that the first two declarations were overridden by the two declarations following them. That's what happens when we override `onunload` and `onload`.

Web APIs

To demonstrate that we can use the Web APIs the exact same way we can use on the browser, we'll write a rudimentary program that fetches the Deno website logo, converts it to base64, and prints to the console an HTML page with the base64 of the image there. Let's do this by following these steps:

1. Start with the request to `https://deno.land/logo.svg`:

    ```
    fetch("https://deno.land/logo.svg")
    ```

2. Convert it into `blob`:

    ```
    fetch("https://deno.land/logo.svg")
      .then(r =>r.blob())
    ```

3. Get the text out of the `blob` object and convert it into `base64`:

    ```
    fetch("https://deno.land/logo.svg ")
      .then(r =>r.blob())
      .then(async (img) => {
        const base64 = btoa(
          await img.text()
        )
      });
    ```

4. Print to the console an HTML page with an image tag using the base64 image:

    ```
    fetch("https://deno.land/logo.svg ")
      .then(r =>r.blob())
      .then(async (img) => {
    ```

```
const base64 = btoa(
    await img.text()
  )

    console.log(`<html>
<img src="data:image/svg+xml;base64,${base64}" />
</html>
    `
  )
})
```

When we run this, we get the expected output:

```
$ deno run --allow-net web-apis/fetch-deno-logo.js
<html>
  <img src="data:image/svg+xml;base64,PHN2ZyBoZWlnaHQ9Ijgx
My4xODQiIHdpZHRoPSI4MTMuMTUiIHhtbG5zPSJodHRwOi8vd3d3Lncz
Lm9yZy8yMDAwL3N2ZyI+PGcgZmlsbD0iIzIyMiI+PHBhdGggZD0ibTM
NC41NzUuMjA5Yy0xLjkuMi04IC45LTEzLjUgMS40LTc4LjIgOC4yLTE1
NS4yIDQxLjMtMjE4IDkzLjktMTEuNiA5LjYtMzggMzYtNDcuNiA0Ny42
LTUyIDYyLjEtODIuNCAxMzEuOC05My42IDIxNC4zLTIuNSAxOC4z
```
...

Now, with the help of *nix output redirection features, we can create an HTML file with the output of our script:

```
$ deno run --allow-net web-apis/fetch-deno-logo.js > web-apis/
deno-logo.html
```

You can now inspect the file, or open it directly in the browser to test that it works.

It is also possible to use your knowledge from the previous chapter and directly run a script from the Deno standard library to serve the current folder:

```
$ deno run --allow-net --allow-read https://deno.land/
std@0.83.0/http/file_server.ts web-apis
Check https://deno.land/std@0.65.0/http/file_server.ts
HTTP server listening on http://0.0.0.0:4507
```

And then, by navigating to `http://localhost:4507/deno-logo.html`, we can check that the image is there and working:

Figure 3.1 – Accessing a web page with the Deno.land logo as a base64 image

Those are just examples of Web APIs that are supported in Deno. In this specific example, we've used `fetch` and `btoa` but more will be used throughout the chapter.

Feel free to experiment with these already familiar APIs, either by writing simple scripts or by using the REPL. In the rest of the book, we'll be using known functions from the Web APIs. In the next section, we'll get to know the Deno namespace, the functions that only work inside Deno, and generally provide a more low-level behavior.

Exploring the Deno namespace

All the functionality that is not covered by a Web API lives under the Deno namespace. This is functionality that is exclusive to Deno and that can't, for instance, be bundled to run in Node or the browser.

In this section, we'll explore some of this functionality. We'll be building small utilities, mimicking some of the programs you use daily.

If you want to explore the available functions before we get our hands dirty, they are available at `https://doc.deno.land/builtin/stable`.

Building a simple ls command

If you've ever used a *nix system's Terminal or Windows PowerShell, you are probably familiar with the `ls` command. Briefly, it lists the files and folders inside a directory. What we will do is create a Deno utility that mimics some functionality of `ls`, that is, lists the files in a directory, and shows some details about them.

The original command has a countless number of flags, which we will not implement here for brevity reasons.

The information we decided to show is the name, size, and last modified date of a file. Let's get our hands dirty:

1. Create a file named `list-file-names.js` and use `Deno.readDir` to get a list of all files and folders in the current directory:

    ```
    for await (const dir of Deno.readDir(".")) {
      console.log(dir.name)
    }
    ```

    This will print the files in the current directory on different lines:

    ```
    $ deno run --allow-read list-file-names.ts
    .vscode
    list-file-names.ts
    ```

    We've used `readDir` (`https://doc.deno.land/builtin/stable#Deno.readDir`) from the Deno namespace.

    As is mentioned in the documentation, it returns `AsyncInterable`, which we're looping through and printing the name of the file. As the runtime is written in TypeScript, we have very useful type completion and we know exactly what properties are present in every `dir` entry.

    Now, we want to get the current directory as a command-line argument.

2. Use `Deno.args` (https://doc.deno.land/builtin/stable#Deno.args) to get the command-line arguments. If no argument is sent, use the current directory as a default:

```
const [path = "."] = Deno.args;
for await (const dir of Deno.readDir(path)) {
    console.log(dir.name)
}
```

We're leveraging array destructuring to get the first value of `Deno.args` and at the same time using default properties to set the default value of the `path` variable.

3. Navigate to the `demo-files` folder (https://github.com/PacktPublishing/Deno-Web-Development/tree/master/Chapter03/ls/demo-files) and run the following command:

```
$ deno run --allow-read ../list-file-names.ts
file-with-no-content.txt
.hidden-file
lorem-ipsum.txt
```

It looks like it is working. It is getting the files from the folder it is currently in and listing them.

We now need to get the file information so that we can display it.

4. Use `Deno.stat` (https://doc.deno.land/builtin/stable#Deno.stat) to get information about the files:

> **Tip**
> If you want to explore how a command behaves, you can run it in debug mode, using `--inspect-brk` or you can try it on the REPL.

```
import { join } from
"https://deno.land/std@0.83.0/path/mod.ts";

const [path = "."] = Deno.args;

for await (const dir of Deno.readDir(path)) {
    let fileInfo = await Deno.stat(join(path, dir.name))
```

```
const modificationTime = fileInfo.mtime;
  const message = [
fileInfo.size.toString().padEnd(4),
`${modificationTime?.getUTCMonth().toString().
padStart(2)}/${modificationTime?.getUTCDay().toString().
padEnd(2)}`,
    dir.name
  ]

  console.log(message.join(""))
}
```

In order to make it clearer, we're using an array to organize our messages here. We're also adding some padding, by using padEnd so that the output is aligned. By running the program we just wrote, while in the Chapter03/ls folder (https://github.com/PacktPublishing/Deno-Web-Development/tree/master/Chapter03/ls/demo-files), we get the following output:

```
$ deno run --allow-read index.ts ./demo-files
12    7/4   .hidden
96    7/4   folder
96    7/4   second-folder
5     7/4   my-best-file
20    7/4   .file1
0     7/4   .hidden-file
```

And we get our list of files and folders in the deno-files directory we sent as a parameter, together with the size in bytes and the creation month and day.

Here, we're using the already known and required --allow-read flag to give Deno permissions to access the filesystem. However, in the previous chapter, we mentioned that there was a different way for Deno programs to ask for permissions, using what we called "dynamic permissions." That's what we'll learn about next.

Using dynamic permissions

When writing Deno programs ourselves, it's very common that we know the required permissions beforehand. However, when writing or executing code that might or might not need some permissions or writing an interactive CLI utility, it might not make sense to ask for all permissions at once. That's what dynamic permissions are for.

Dynamic permissions allow programs to ask for permissions as they are needed, allowing whoever is executing the code to give or deny specific permissions interactively.

This is a feature that is still unstable and thus its APIs can change, but I think it's still worth mentioning, because of the amount of potential it enables.

You can have a look at Deno's permissions API at `https://doc.deno.land/builtin/unstable#Deno.permissions`.

What we'll do next is make sure that our `ls` program asks for filesystem read permissions. Let's do it by following these steps:

1. Use `Deno.permissions.request` to ask for read permissions before executing the program:

   ```
   ...
   const [path = "."] = Deno.args;

   await Deno.permissions.request({
     name: "read",
     path,
   });

   for await (const dir of Deno.readDir(path)) {
   ...
   ```

   This asks for permissions for the directory where the program is going to run.

2. Run the program and grant permissions on the current directory:

   ```
   $ deno run --unstable list-file-names-interactive-
   permissions.ts .
   Deno requests read access to ".". Grant? [g/d (g = grant,
   d = deny)] g
   list-file-names-color.ts
   list-file-names.ts
   demo-files
   list-file-names-interactive-permissions.ts
   ```

   By answering g to the permission request command, we're granting it access to the current directory (.).

We can now try to run the same program but denying the permissions this time.

3. Run the program and deny read permissions on the current directory:

```
$ deno run --unstable list-file-names-interactive-
permissions.ts .
Deno requests read access to ".". Grant? [g/d (g = grant,
d = deny)] d
error: Uncaught (in promise) PermissionDenied: read
access to ".", run again with the --allow-read flag
    at processResponse (deno:core/core.js:223:11)
    at Object.jsonOpAsync (deno:core/core.js:240:12)
    at async Object.[Symbol.asyncIterator] (deno:cli/
rt/30_fs.js:125:16)
    at async list-file-names-interactive-permissions.
ts:10:18
```

And that's how dynamic permissions work!

Here, we've used them to control the filesystem read permissions, but they can be used to ask for access to all the available permissions (mentioned in *Chapter 2, The Toolchain*) in the runtime. They're very useful when writing CLI applications, allowing you to interactively tune which permissions the running program has access to.

Using the filesystem APIs

Accessing the filesystem is one of the basic needs we have when writing programs. As you have probably already seen in the documentation, Deno provides APIs to do these common tasks.

With a decision to standardize communication with the Rust core, all of these APIs return Uint8Array and the decoding and encoding should be made by their consumers. This is a substantial difference from Node.js, where some functions return converted formats, where others return blobs, buffers, and so on.

Let's explore these filesystem APIs and read the contents of a file.

We're going to read the example file available at https://github.com/ PacktPublishing/Deno-Web-Development/tree/master/Chapter03/ file-system/sentence.txt, using the TextDecoder and Deno.readFile APIs, as the following script demonstrates:

```
const decoder = new TextDecoder()
const content = await Deno.readFile('./sentence.txt');


console.log(decoder.decode(content))
```

You can note that we've used the `TextDecoder` class, another API that is present in the browser.

Do not forget to use the `--allow-read` permission when running the script so it can read from the filesystem.

If we want to write the content of this file to another file, we can use `writeFile`:

```
const content = await Deno.readFile("./sentence.txt");
await Deno.writeFile("./copied-sentence.txt", content)
```

Note that we don't need the `TextEncoder` anymore since we're using `Uint8Array` we got from `readFile` to send directly to the `writeFile` method. Remember to use the `--allow-write` flag when running it, since it's now writing to the filesystem.

As you probably guessed or read in the documentation, Deno provides an API exactly for that, `copyFile`:

```
await Deno.copyFile("./copied-sentence.txt",
    "./using-copy-command.txt");
```

Now, you probably noticed we're always using `await` before the method calls on Deno namespace functions.

All asynchronous operations on Deno return a promise, and that's the main reason we're doing this. We could use the equivalent `then` syntax and deal with the result there, but we find this to be more readable.

Other APIs for removing, renaming, changing permissions, and so on are also included in the Deno namespace, as you can find in the documentation.

> **Important note**
> Many of the asynchronous APIs in Deno have an equivalent *synchronous* API that can be used for specific use cases where you want to block the process and get a result (for example, `readFileSync`, `writeFileSync`, and so on).

Using buffers

Buffers represent regions in memory that are used to store temporary binary data. They are commonly used to deal with I/O and network operations. As asynchronous operations are something where Deno excels, we'll be exploring buffers in this section.

Deno buffers differ from Node buffers. This happens because when Node was created, and up until version 4, there was no support in JavaScript for `ArrayBuffers`. As Node optimized for asynchronous operations (where buffers really shine), the team behind it had to create a Node buffer to emulate the behavior of a native buffer. Later, `ArrayBuffers` were added into the language and the Node team migrated the existing buffer to leverage it. It currently isn't more than a subclass of `ArrayBuffers`. This same buffer was then deprecated in v10 of Node. As Deno was recently created, its buffer deeply leverages `ArrayBuffer`.

Reading and writing from Deno.Buffer

Deno provides a dynamic length buffer that is implemented on top of `ArrayBuffer`, a fixed memory allocation. Buffers provide functionality similar to a queue where data can be written and read by different consumers. As we initially mentioned, they are heavily used for jobs such as networking and I/O as they allow asynchronous reading and writing.

To give an example, imagine you have an application that is writing some logs that you want to process. You can do it synchronously as they come, or you can have that application writing to a buffer and have a consumer processing them asynchronously.

Let's write a small program for that situation. We will write two short programs. The first one will emulate an application producing logs; the second will consume those logs by using a buffer.

We'll start by writing code that emulates an application producing logs. At `https://github.com/PacktPublishing/Deno-Web-Development/blob/master/Chapter03/buffers/logs/example-log.txt`, there's a file that has some example logs we'll use:

```
const encoder = new TextEncoder();

const fileContents = await Deno.readFile("./example-log.txt ");

const decoder = new TextDecoder();
const logLines = decoder.decode(fileContents).split("\n");

export default function start(buffer: Deno.Buffer) {
  setInterval(() => {
    const randomLine = Math.floor(Math.min(Math.random() *
      1000, logLines.length));
    buffer.write(encoder.encode(logLines[randomLine]));
```

```
    },    100)
}
```

This code reads the content from the example file and splits it into lines. Then, it gets a random line number and every 100 ms writes that line into a buffer. This file then exports a function that we can call to start "generating random logs." We'll use this in the next script to mimic an application producing logs.

Now comes the interesting part: we'll write our basic *log processor* by following these steps:

1. Create a buffer and send it to the start function of the log producer we just wrote:

   ```
   import start from "./logCreator.ts";

   const buffer = new Deno.Buffer();

   start(buffer);
   ```

2. Call the processLogs function to start processing the log entries present in the buffers:

   ```
   ...
   start(buffer);
   processLogs();

   async function processLogs() {}
   ```

   As you can see, the processLogs function would be called and nothing would happen, as we haven't implemented a program to do it yet.

3. Create an object type of Uint8Array inside the processLogs function and read the content of the buffer there:

   ```
   ...
   async function processLogs() {
     const destination = new Uint8Array(100);
     const readBytes = await buffer.read(destination);
     if (readBytes) {
       // Something was read from the buffer

     }
   }
   ```

The documentation (`https://doc.deno.land/builtin/stable#Deno.Buffer`) states that when there is something to read, the `read` function from `Deno.Buffer` returns the number of bytes read. When there is nothing to read, the buffer is empty and it returns null.

4. Now, inside `if`, we can just decode the content that was read, as we know it comes in `Uint8Array` format:

```
const decoder = new TextDecoder();
...
if (readBytes) {
  const read = decoder.decode(destination);
}
```

5. To print the decoded value on the console, we can use the already known `console.log`. We can also do it differently, by using `Deno.stdout` (`https://doc.deno.land/builtin/stable#Deno.stdout`) to write to the standard output.

`Deno.stdout` is a `writer` object in Deno (`https://doc.deno.land/builtin/stable#Deno.Writer`). We can use its `write` method to send text there:

```
const decoder = new TextDecoder();
const encoder = new TextEncoder();
...
if (readBytes) {
  const read = decoder.decode(destination);
  await Deno.stdout.write(encoder.encode(`${read}\n`));
}
```

And with this, we're writing to `Deno.stdout`, the value we just read. We're also adding a line break at the end (`\n`) so that it becomes a little more readable on the console.

If we leave it this way, this `processLogs` function will run only once. As we want this to run again and check if there are more logs in `buffer`, we'll need to schedule it to run again later.

6. Use `setTimeout` to call the same `processLogs` function 100 ms from now:

```
async function processLogs() {
    const destination = new Uint8Array(100);
    const readBytes = await buffer.read(destination);
    if (readBytes) {
        ...
    }

    setTimeout(processLogs, 10);
}
```

As an example, if we open the `example-log.txt` file, we can see that there are lines that contain dates in the following format: `Thu Aug 20 22:14:31 WEST 2020`.

Let's imagine we just want to print logs that have `Tue` on them. Let's write the logic to do that:

```
async function processLogs() {
  const destination = new Uint8Array(100);
  const readBytes = await buffer.read(destination);
  if (readBytes) {
    const read = decoder.decode(destination);

    if (read.includes("Tue")) {
      await Deno.stdout.write(encoder.encode(`${read}\n`));
    }
  }

  setTimeout(processLogs, 10);
}
```

Then, we execute the program while inside the folder containing the `example-logs.txt` file:

```
$ deno run --allow-read index.ts
Tue Aug 20 17:12:05 WEST 2019
Tue Sep 17 02:19:56 WEST 2019
Tue Dec  3 14:02:01 CET 2019
Tue Jul 21 10:37:26 WEST 2020
```

The log lines with the dates appear as they are read from the buffer and match our criteria.

This was a short demonstration of what can be done with buffers. We were able to asynchronously write and read from a buffer. This approach allows, for instance, a consumer to be working on a portion of a file while the application is reading other parts of it.

The Deno namespace provides a lot more functionality than what we've tried here. In this section, we decided to pick a few parts and give you a glimpse of how much it enables.

We'll be using these functions, together with third-party modules and the standard library, when we write our web server, from *Chapter 4, Building a Web Application*, onwards.

Using the standard library

In this section, we'll explore the behavior provided by Deno's standard library. It is currently not considered stable by the runtime and thus modules are separately versioned. At the time we're writing, the standard library is at *version 0.83.0*.

As we previously mentioned, Deno is very meticulous in what it adds to the standard library. The core team wants it to provide enough behavior, so people don't need to rely on millions of external packages to do certain things, but at the same time doesn't want to add too much of an API surface. This is a fine balance that is hard to strike.

With the assumed inspiration of golang, most of the Deno standard library functions mimic the language created by Google. This happens because the Deno team truly believes in the way *golang* evolved its standard library, one that is commonly known for being well polished. As a funny note, Ryan Dahl (Deno and Node creator) mentions in one of his talks that, when pull requests add new APIs to the standard library, the equivalent *golang* implementation is asked for.

We'll not go over the whole library for the same reasons we didn't go over the whole Deno namespace. What we'll do is build a couple of useful programs with it while we learn what it enables. We'll go from stuff such as generating IDs, to logging, to HTTP communication, among other known use cases.

Adding colors to our simple ls

A few pages ago, we built a very rough and simple "clone" of the `ls` command in *nix systems. At the time we listed the files, together with their size and modification date.

To start exploring the standard library, we're going to add some coloring to the terminal output of that program. Let's make folder names be printed in red so we can easily differentiate them.

We'll create a file called `list-file-names-color.ts`. This time we will be using TypeScript as we'll get much better completion because the standard library and the Deno namespace functions were written with that.

Let's explore the standard library functions that allow us to colorize our text (`https://deno.land/std@0.83.0/fmt/colors.ts`).

If we want to look at a module's documentation, we can go directly to code, but we can also use the `doc` command or the documentation website. We'll use the latter.

Navigate to `https://doc.deno.land/https/deno.land/std@0.83.0/fmt/colors.ts`. All the listed available methods are presented on the screen:

1. Import the method from the standard library's formatting library that prints the text in red:

    ```
    import { red } from "https://deno.land/std@0.83.0/fmt/
    colors.ts";
    ```

2. Use it inside our `async` iterator that is going through our files in the current directory:

```
const [path = "."] = Deno.args;
for await (const item of Deno.readDir(path)) {
  if (item.isDirectory) {
    console.log(red(item.name));
  } else {
    console.log(item.name);
  }
}
```

3. By running it inside the `demo-files` folder (`https://github.com/ PacktPublishing/Deno-Web-Development/tree/master/ Chapter03/ls`), we get the folders printed in red (it is not possible to see this in the printed book, but you can run it locally):

```
$ deno run -allow-read list-file-names-color.ts
file-with-no-content.txt
demo-folder
.hidden-file
lorem-ipsum.txt
```

We now have a better `ls` command that enables us to distinguish folders from files, using the coloring functions from the standard library. There are many other modules provided by the standard library that we'll have a look at during the course of the book. Some of them will be used when we start writing our own application.

One module that we'll pay special attention to is the HTTP module, which we'll heavily use from the next section onwards.

Building a web server using the HTTP module

The main focus of this book, together with presenting Deno and how it can be used, is to learn how to use it to build web applications. Here, we'll create a simple JSON API to introduce you to the HTTP module.

We'll build an API that will save and list notes. We will call these notes post-its. Imagine that this is the API that will feed your post-its board.

We'll create a very simple routing system with the help of Web APIs and the functions from the Deno standard library's HTTP module. Keep in mind we're doing this to explore the APIs themselves and thus this is not production-ready code.

Let's start by creating a folder named `post-it-api` and a file named `index.ts` inside. One more time, we will use TypeScript as we believe the autocomplete and type checking capabilities greatly improve our experience and reduce the number of possible errors.

The final code for this section is available at `https://github.com/ PacktPublishing/Deno-Web-Development/blob/master/Chapter03/ post-it-api/steps/7.ts`:

1. Start by importing the standard library HTTP module into our file:

    ```
    import { serve } from
        "https://deno.land/std@0.83.0/http/server.ts";
    ```

2. Write the logic to handle requests by using `AsyncIterator`, as we did in previous examples:

    ```
    console.log("Server running at port 8080");
    for await (const req of serve({ port: 8080 })) {
      req.respond({ body: "post-it api", status: 200 });
    }
    ```

    If we now run it, this is what we get. Keep in mind we need to use the `--allow-net` flag, mentioned in the Permissions section, for it to have network access:

    ```
    deno run --allow-net index.ts
    Server running at port 8080
    ```

3. For clarity, we can extract the port and the server instance to a separate variable:

    ```
    const PORT = 8080;
    const server = serve({ port: PORT });


    console.log("Server running at port", PORT);
    for await (const req of serve({ port: PORT })) {
      ...
    ```

And we have our server working, as it was before, with the small difference that now the code looks (arguably) more readable with the configuration variables at the top of the file. We'll later learn how can we extract those from the code.

Returning a list of post-its

Our first requisite is that we have an API that returns a list of post-its. Those will be composed of the name, title, and the date created. Before we get there, and to enable us to have multiple routes, we need a routing system.

For the purpose of this exercise, we'll build ours. This is our way of getting to know some of the APIs built into Deno. We'll later agree that when writing production applications, it is sometimes better to reuse tested and heavily used pieces of software than to keep reinventing the wheel. However, it is completely fine to *reinvent the wheel* for learning purposes.

To create our basic routing system, we will use some APIs that you probably know from the browser. Objects such as URL, UrlSearchParams, and so on.

Our goal is to be able to define a route by its URL and path. Something like GET /api/post-its would be nice. Let's do it!

1. Start by creating a URL object (https://developer.mozilla.org/en-US/docs/Web/API/URL) to help us parse the URL and its parameters. We'll extract HOST and PROTOCOL to a different variable, so we don't have to repeat ourselves:

    ```
    const PORT = 8080;
    const HOST = "localhost";
    const PROTOCOL = "http";


    const server = serve({ port: PORT, hostname: HOST });

    console.log(`Server running at ${HOST}:${PORT}`);
    for await (const req of server) {
      const url = new
        URL(`${PROTOCOL}://${HOST}${req.url}`);
      req.respond({ body: "post-it api", status: 200 });
    }
    ```

2. Use the created URL object to do some routing. We'll use a switch case for that. When no route matches, a 404 should be sent to the client:

    ```
    const pathWithMethod = `${req.method} ${url.pathname}`;
    switch (pathWithMethod) {
      case "GET /api/post-its":
        req.respond({ body: "list of all the post-its",
          status: 200 });
    ```

```
        continue;
    default:
        req.respond({ status: 404 });
    }
```

> **Tip**
>
> You can use the `--unstable` and `--watch` flags together when running your script to restart it on file changes as follows: `deno run --allow-net --watch --unstable index.ts`.

3. Access `http://localhost:8080/api/post-its` and confirm we have the correct response. Any other routes will get a 404 response.

   Note that we're using the `continue` keyword to make Deno jump out of the current iteration after responding to the request (remember we're inside a `for` loop).

   You might have noticed that, at the moment, we're just routing by path and not by method. This means any request made to `/api/post-its`, either POST or GET, will get the same response. Let's fix that by moving ahead.

4. Create a variable that contains the request method and the pathname:

   ```
   const pathWithMethod = `${req.method} ${url.pathname}`
   switch (pathWithMethod) {
   ```

   We can now define our routes the way we desire, GET `/api/post-its`. Now that we have the basics of our routing system, we'll write the logic to return our post-its.

5. Create the TypeScript interface that will help us maintain the structure of the post-its:

   ```
   interface PostIt {
       title: string,
       id: string,
       body: string,
       createdAt: Date
   }
   ```

6. Create a variable that will work as our *in-memory database* for this exercise.

    We'll use a JavaScript object where the keys are the IDs and values are the objects of the `PostIt` type we just defined:

    ```
    let postIts: Record<PostIt["id"], PostIt> = {}
    ```

7. Add a couple of fixtures to our database:

    ```
    let postIts: Record<PostIt["id"], PostIt> = {
        '3209ebc7-b3b4-4555-88b1-b64b33d507ab': { title: 'Read
    more', body: 'PacktPub books', id: 3209ebc7-b3b4-4555-
    88b1-b64b33d507ab ', createdAt: new Date() },
        'a1afee4a-b078-4eff-8ca6-06b3722eee2c': { title:
    'Finish book', body: 'Deno Web Development', id:
    '3209ebc7-b3b4-4555-88b1-b64b33d507ab ', createdAt: new
    Date() }
    }
    ```

    Note that we're *generating* the *IDs* by hand for now. Later, we'll use another module from the standard library to do it. Let's get back to our API and change the `case` that handles our route.

8. Change the `case` that will return all the post-its instead of the hardcoded message.

    As our database is a key/value store, we need to use `reduce` to build an array with all our post-its (delete the line highlighted in the code block):

    ```
    case GET "/api/post-its":
      req.respond({ body: "list of all the post-its", status:
        200 });
      const allPostIts = Object.keys(postIts).
        reduce((allPostIts: PostIt[], postItId) => {
          return allPostIts.concat(postIts[postItId]);
        }, []);

      req.respond({ body: JSON.stringify({ postIts:
        allPostIts }) });
      continue;
    ```

9. Run the code and go to /api/post-its. We should have our post-its listed there!

    You might have noticed that it is still not 100 percent correct, since our API is returning JSON, and its headers do not match the payload.

10. We'll add the content-type by using an API we know from the browser, the Headers object (https://developer.mozilla.org/en-US/docs/Web/API/Headers). Delete the line highlighted in the following code block:

```
const headers = new Headers();
headers.set("content-type", "application/json");

const pathWithMethod = `${req.method} ${url.pathname}`
switch (pathWithMethod) {
  case "GET /api/post-its":
  ...
      req.respond({ body: JSON.stringify({ postIts:
        allPostIts }) });
      req.respond({ headers, body: JSON.stringify({
        postIts: allPostIts }) });
      continue;
```

We've created an instance of the Headers object up there, and then we used it on the response, on req.respond. This way, our API is now more coherent, digestible, and following standards.

Adding a post-it to the database

Now that we have a way to read our post-its, we will need a way to add new ones as it doesn't make much sense to have an API with completely static content. That's what we'll do.

We'll use the *routing infrastructure* we created to add a route that allows us to *insert* records into our database. Since we're following REST guidelines, that route will live on the same path as the one that lists post-its, but with a different method:

1. Define a route that always returns the 201 status code:

```
case "POST /api/post-its":
  req.respond({ status: 201 });
  continue
```

2. Testing it, with the help of `curl`, we can see it's returning the correct status code:

```
$ curl -I -X POST http://localhost:8080/api/post-its
HTTP/1.1    201 Created
content-length: 0
```

> **Important note**
> We're using `curl` but feel free to use your favorite HTTP requests tool, you can even use a graphical client such as Postman (`https://www.postman.com/`).

Let's make the new route do what it is supposed to. It should get a JSON payload and use that to create a new post-it.

We know, by looking at the documentation of the standard library's HTTP module (`https://doc.deno.land/https/deno.land/std@0.83.0/http/server.ts#ServerRequest`) that the body of the request is a *Reader* object. The documentation includes an example on how to read from it.

3. Following the recommendation, read the value and print it to get a better understanding of it:

```
case "POST /api/post-its":
    const body = await Deno.readAll(req.body);
    console.log(body)
```

4. Make a request with `body`, with the help of `curl`:

```
$ curl -X POST -d "{\"title\": \"Buy milk\"}" http://localhost:8080/api/post-its
```

The request succeeds but it doesn't return anything more than a `201` status code. If we look at our running server though, something like this is printed to the console:

```
Uint8Array(25) [
    123,  34, 116, 105, 116, 108, 101,
    34,58,32,34,84,  101, 115,
    116,  32, 112, 111, 115, 116,  45,
    105, 116,  34, 125
]
```

We previously learned that Deno uses `Uint8Array` to do all its communications with the Rust backend, and this is not an exception. However, `Uint8Array` is not what we currently want, we want the actual text of the request body.

5. Use `TextDecoder` to get the request body as a readable value. After doing this, we'll log the output again and we'll make a new request:

```
$ deno -X POST -d "{\"title\": \"Buy milk\"}"
http://localhost:8080/api/post-its
```

This is what the server printed to the console this time:

```
{"title": "Buy milk "}
```

We're getting there!

6. Since the body is a string, we need to parse it into a JavaScript object. We'll use an old friend of ours, `JSON.parse`:

```
const decoded = JSON.parse(new
  TextDecoder().decode(body));
```

We now have the request body in a format we can act on, and that's pretty much all it takes for us to create a new database record. Let's create one by following these steps:

7. Use the `uuid` module (`https://deno.land/std@0.83.0/uuid`) from the standard library to generate a random UUID for our records:

```
import { v4 } from
  "https://deno.land/std/uuid/mod.ts";
```

8. In our route's switch case, we'll create an `id` with the help of the `generate` method and insert it in the *database*, adding the `createdAt` date on top of what the user sent in the request payload. For the sake of this example, we're skipping validation:

```
case "POST /api/post-its":
...
    const decoded = JSON.parse(new
      TextDecoder().decode(body));

    const id = v4.generate();
    postIts[id] = {
      ...decoded,
      id,
```

```
    createAt: new Date()
  }
  req.respond({ status: 201, body:
    JSON.stringify(postIts[id]), headers });
```

Note that we're using the same `headers` object we previously defined (in the `GET` route) so that our API responds with `Content-Type: application/json`.

Then again, as we follow the *REST* guidelines, we return the `201 Created` code and the created record.

9. Save the code, restart the server, and run it again:

```
$ curl -X POST -d "{\"title\": \"Buy groceries\",
\"body\":\"1 x Milk\"}" http://localhost:8080/post-
its
{"title":"Buy    groceries","body":"1 x
Milk","id":"9a3c6a56-713b-4b8c-80a0-
8d5a37aaefee","createdAt":"2020-09-08T00:13:46.124Z"}
```

And we have it working! We can now do a `GET` request to the route that lists all the post-its to check if the record was actually inserted into the database:

```
$ curl http://localhost:8080/api/post-its
{"postIts":[{"title":"Read more","body":"PacktPub
books","id":"3209ebc7-b3b4-4555-88b1-
b64b33d507ab","createdAt":"2021-01-10T16:28:5
2.210Z"},{"title":"Finish book","body":"Deno
Web Development","id":"a1afee4a-b078-4eff-8ca6-
06b3722eee2c","createdAt":"2021-01-10T16:28:52.210Z"},{"
title":"Buy groceries","body":"1 x Milk","id":"b35b0a62-
4519-4491-9ba9-b5809b4810d5","createdAt":"2021-01-
10T16:29:05.519Z"}]}
```

And it works! We now have an API that returns and adds post-its to a list.

This pretty much wraps up what we'll do in terms of APIs with the HTTP module for this chapter. As most of the APIs, like the one we wrote, are made to be consumed by a frontend application, we'll do that to wrap up this chapter.

Serving a frontend

Since it is out of the scope of this book, we will not write the frontend code that interacts with this API. However, if you want to use it with a single-page application that fetches the post-its and displays them, I've included one in the book's files (`https://github.com/PacktPublishing/Deno-Web-Development/blob/master/Chapter03/post-it-api/index.html`).

What we'll learn here is how can we use the web server we just built to serve an HTML file:

1. First, we need to create a route at the root of our server. Then, we need to set the correct `Content-Type` and return the file's content by using the already known filesystem APIs.

    In order to get the path to the HTML file in reference to the current file, we'll use the URL object together with the `import.meta` declaration from JavaScript (`https://developer.mozilla.org/en-US/docs/Web/JavaScript/Reference/Statements/import.meta`), which contains the path to the current file:

    ```
    import { resolve, fromFileUrl } from
    "https://deno.land/std@0.83.0/path/mod.ts";
    ...
        case "GET /":
            const file = await Deno.readFile(
                resolve(fromFileUrl(import.meta.url), "..",
                    "index.html")
            );
            let htmlHeaders = new Headers();
            htmlHeaders.set("content-type", "text/html");

            req.respond({ body: new TextDecoder().decode(file),
                headers: htmlHeaders })
            continue;
    ```

We're using the `resolve`, and `fromFileUrl` methods from Deno's standard-library to get a URL that is relative to the current file.

Note that we now need to run this with the `--allow-read` flag since our code is reading from the filesystem.

2. In order for us to be a little more secure, we will specify the exact folder the program can read, by sending it to the `--allow-read` flag:

```
$ deno run --allow-net --allow-read=. index.ts
Server running at http://0.0.0.0:8080
```

This will prevent us from any bugs that might allow malicious people to read from our filesystem.

3. Access the URL with the browser, and you should get to a page where we can see the fixture `post-its` we've added. To add a new one, you can also click the **Add a new post-it** text and fill in the form:

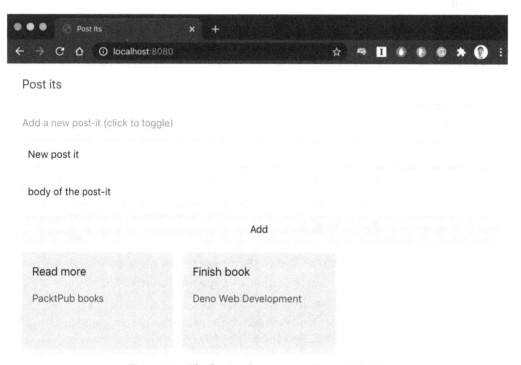

Figure 3.2 – The frontend consuming the post-it API

> **Important note**
>
> Please keep in mind that, in many production environments, it is not a recommended practice to have your API serving your frontend code. Here, we did it for learning purposes, so we could understand some of the possibilities of the standard library HTTP module.

In this section, we learned how can we use the modules provided by the standard library to our benefit. We made a simple version of a very common command, `ls`, and used the output formatting functions from the standard library to add some color to it. To finish the section, we made an HTTP API with a couple of endpoints that listed and persisted records. We went through different requirements and learned how Deno can be used to accomplish them.

Summary

As we go through the book, our knowledge of Deno gets more practical and we start to use it for use cases that are closer to the real world. That was what this chapter was about.

We started the chapter by learning about some fundamental characteristics of the runtime, namely the program lifecycle, and how Deno sees module stability and versioning. We rapidly moved on to the Web APIs provided by Deno by writing a simple program that fetches the Deno logo from the website, converts it to base64, and puts it into an HTML page.

Then, we got into the `Deno` namespace and explored some of its low-level functionality. We built a couple of examples with the filesystem API and ended up building a rudimentary copy of the `ls` command with it.

Buffers are things that are heavily used in the Node world, with their capabilities to perform asynchronous read and write behavior. As we know, Deno shares many use cases with Node.js, and that made it impossible to not talk about buffers in this chapter. We started by explaining how Deno buffers differ from Node.js' and ended the section by building a small application that handles reading and writing asynchronously from them.

To wrap up the chapter, we got closer to one of the main goals of this book, using Deno for web development. We created our first JSON API with Deno. In the process, we learned about multiple Deno APIs and we even built our basic routing system. We then proceeded by creating a couple of routes that listed and created records in our *data storage*. Getting closer to the end of this chapter, we learned how we can handle headers in our APIs and added those to our endpoints.

We finished the chapter by serving a single-page application directly from our web server; that same single-page application that consumed and interacted with our API.

This was a chapter where lots of ground was covered. We started building APIs that are now much closer to reality than what we previously did. We also got more of a grasp on what it is to develop with Deno, using permissions, and the documentation.

The current chapter wraps up our introductory journey and has hopefully left you curious about what is coming next.

In the next four chapters, we'll be building a web application and will explore all the decisions made in the process. Most of the knowledge you have learned so far will be used later, but there is also a ton of new, exciting stuff coming up. In the next chapter, we'll start creating an API that we'll be adding features to as the chapters progress.

I hope to have you on board!

Section 2: Building an Application

In this hands-on section, you will create a Deno application, starting from a server-side rendered website and then moving on to **Representational State Transfer (REST) application programming interfaces (APIs)** that connect with databases and have authentication.

This section contains the following chapters:

4

Building a Web Application

Here we are! We have traveled down a long road to get here. This is where all the fun starts. We've been through three phases: getting to know what Deno is, exploring the toolchain provided by it, and understanding the details and functionality available via its runtime.

Pretty much all the content from the previous chapters will prove to be useful in this one. Hopefully, the introductory chapters made you feel confident enough to start applying what we have learned together. We'll use those chapters, together with your existing TypeScript and JavaScript knowledge, to build a complete web application.

We'll be writing an API that contains business logic, handles authentication, authorization, and logging, and much more. We will cover enough of the fundamental pieces for you to, at the end, feel comfortable choosing Deno to build your next great app.

In this chapter, instead of talking just about Deno, we will also go over some thoughts regarding the fundamentals of software engineering and application architecture. We believe it is crucial to keep some things in mind when it comes to building an application from scratch. We will look at of the fundamentals, which will be proven useful and help us structure our code, enabling it to evolve in the future by making it easy to change.

Later, we will start to reference some third-party modules, look at their approaches, and decide on what we'll use from here on to help us deal with routing and HTTP related challenges. We'll also make sure that we structure our application in a way that the third-party code is isolated and works as an enabler for the functionalities we want to build, more than the functionalities themselves.

We will cover the following topics in this chapter:

- Structuring a web application
- Exploring Deno HTTP frameworks
- Let's get started!

Technical requirements

The code files used in this chapter are available at the following link: `https://github.com/PacktPublishing/Deno-Web-Development/tree/master/Chapter04/museums-api`.

Structuring a web application

When starting an application, it's important that we spend some time thinking about its structure and architecture. That's where there this section will start: by looking at the backbone of application architecture. We'll have a look at what advantages it brings and align ourselves with a set of principles that will help us scale it as the application grows.

Then, we'll develop what will become the application's first endpoint. However, first, we'll start with the business logic. The persistency layer will follow, and we'll finish by looking at an HTTP endpoint that will act as the application's entry point.

Deno as an unopinionated tool

When we're using tools that are low level and delegate many decisions to developers, such as Node.js and Deno, structuring an application is one of the big challenges that arises.

This is very different compared to an opinionated web framework, such as PHP Symfony, Java SpringBoot, or Ruby on Rails, where many of these decisions are made for us.

Most of these decisions have something to do with structure; that is, code and folder structure. Those frameworks normally provide us with ways to deal with dependencies, imports, and even provide some guidance regarding different application layers. Since we're using the *raw* language with a few packages, we will take care of structure by ourselves in this book.

The aforementioned frameworks can't be directly compared with Deno since they are frameworks built on top of languages, such as PHP, Java, and Ruby. But when we look at the JS world, namely at Node.js we can observe that the most popular tools used to create HTTP servers are Express.js and Kao. These tend to be much lighter than the aforementioned frameworks, and even though there are also some solid complete alternatives such as Nest.js or hapi.js, the Node.js community tends to prefer a *library* approach more than a *framework* one.

Even though these very popular libraries provide a good amount of functionality, many decisions are still delegated to developers. This isn't the libraries' fault, but more a community preference.

On one hand, having direct access to these primitives lets us build applications that are very well-suited to our use cases. On the other hand, flexibility is a trade-off. Having a lot of flexibility comes the responsibility of making an innumerous number of decisions. And when it comes to making many decisions there are many opportunities to make bad decisions. The hard part is that these are normally decisions that drastically influence the way a code base scales, and that's what gives them such importance.

In its current state, Deno and its community are following an approach that's very similar to Node.js on this framework-versus-library subject. The community is mostly betting on light and small pieces of software that are created by developers to fit their specific needs. We'll evaluate some of these later in this chapter.

Starting here, and throughout this book, we'll use an application structure that we believe offers great benefits for the use case at hand. However, don't expect that structure and architecture to be a silver bullet, as we are pretty sure such things do not exist in the software world; every architecture will have to keep evolving as it grows.

Instead of just throwing in a recipe and following it, we want to become familiar with a way of thinking – a rationale. This should enable us to make correct decisions further down the road with one objective in mind: *writing code that is easy to change*.

By writing code that is easy to change, we're always ready to change our application for the better without much effort.

The most important part of an application

Applications are created to fit a purpose. It doesn't matter if that purpose is to support a business or a simple pet project. At the end of the day, we want it to do something. That *something* is what makes the application useful.

This might seem obvious, but it is sometimes very easy for us, as developers, to get so enthusiastic about a technology that we forget that it is just a means to an end.

As Uncle Bob says in his *Architecture – the lost years* talk (`https://www.youtube.com/watch?v=hALFGQNeEnU`), it is very common for people to forget the application's purpose and focus more on the technology itself. It is very important that we remember this in all the phases of application development, but it is even more critical when we're setting up its initial structure. Next, we'll discover the requirements for the application we'll be building throughout the remainder of this book.

What is our application about?

Even though we truly believe that business logic is the most important thing in any application, in this book, the case is a little different. We'll be creating an example application, but it will just be a means to reach the main goal: learning Deno. However, as we want the process to be as real as possible, we want to have a clear objective in mind.

We will build an application that will let people create and interact with a list of museums. We can make this clearer by listing its features as user stories, as follows:

- The user is able to register and log in.

- The user is able to create a museum with a title, description, and location.

- The user can view a list of museums.

Throughout this journey, we'll develop APIs and the logic to support those features.

Now that we're familiar with the end goals, we can start thinking about how to structure the application.

Understanding folder structure and application architecture

The first thing we need to be aware of regarding the folder structure, especially when we're starting a project from scratch without a framework, is that it will keep evolving with the project. A folder structure that is good for a project with a couple of endpoints will not be as good for a project with hundreds of them. This depends on many things, from team size, to the standards defined, and ultimately to preferences.

When defining the folder structure, it is important that we get to a place where we can facilitate future decisions about where to locate a piece of code. The folder structure should provide clear hints on how to make good architectural decisions.

At the same time, we certainly don't want to create an overengineered application. We'll create enough abstractions so that modules are very contained and do not have knowledge outside their domain, but not more than that. Keeping this in mind also forces us to build flexible code and clear interfaces.

Ultimately, what's most important is that the architecture enables the code base to be as follows:

- Testable

- Easy to extend

- Decoupled from a specific technology or library

- Easy to navigate and reason about

We'll have to keep in mind that, while creating folders, files, and modules, we don't want any of the previously listed topics to be compromised.

These principles are very much in line with the SOLID principles of software design, made famous by Uncle Bob, Robert C. Martin (`https://en.wikipedia.org/wiki/SOLID`) in another talk worth watching (`https://youtu.be/zHiWqnTWsn4`).

The folder structure we are going to use in this book might sound familiar to you if you come from a Node.js background.

As it also happens with Node.js there's nothing preventing us from creating a full API in a single file. However, we will not do this as we believe that some initial separation of concerns will vastly improve our flexibility later, without sacrificing the developer's productivity.

In the following section, we'll look at the responsibilities of the different layers and how they fit together while developing a feature for our application.

By following this line of thought, we strive to guarantee a degree of decoupling between modules. For instance, we want to ensure that making a change in the web framework doesn't means we have to touch business logic objects.

All these recommendations, as well as the ones we will be making throughout this book, will help ensure that the central part of our application is our business logic, with everything else being just plugins. A JSON API is just a way of sending our data to our users, while a database is just a way to persist data; neither of these should be central parts of the application.

One way to make sure we're doing this is to do the following mental exercise when we are writing the code:

"When you're writing business logic, imagine that these objects will be used in a different context. Take, for instance, using the business logic with a different delivery mechanism (a CLI, for instance) or a different persistence engine (an in-memory database instead of a NOSQL database)."

In the next few pages, we'll walk you through how to create different layers, and we'll also explain all the design decisions and what they enable.

Let's get practical and start creating our project's backbone.

Defining the folder structure

The first thing we'll do in our project's folder is create an `src` folder.

This is, predictably, where our code will live. We don't want any code to be at the root level of the project because configuration files, READMEs, documentation folders, and so on might be added there. This would make it hard to distinguish the code.

We'll be spending most of our time inside the `src` folder in the following chapters. Since our application is about museums, we'll create a folder inside the `src` folder named `museums`. This is where most of the logic that will be written in this chapter will live. Later, we'll create files for types, controllers, and repositories. Then, we'll create the `src/web` folder.

The controller's file is where our business logic will live. The repository will take care of the logic related to data access, while the web layer will handle everything that is *web-related*.

You can see what the final structure will look like by taking a look at the GitHub repository of this book: `https://github.com/PacktPublishing/Deno-Web-Development/tree/master/Chapter04/museums-api`.

The initial requirement for this chapter is that there is a route where we can perform a GET request and receive a list of museums in JSON format.

We will start in the controller's file (`src/museums/controller.ts`) and write the required business logic.

This is how the folder's structure should look like:

```
└── src
    ├── museums
    │   ├── controller.ts
    │   ├── repository.ts
    │   └── types.ts
    └── web
```

This is our starting point. Everything that is related to museums will be located inside the museums folder, and we'll call it a module. The controller file will host the business logic, the repository file will host data fetching capabilities, and the types file will be where our types will be located.

Now, let's start coding!

Developing the business logic

We previously stated that our business logic is the most important part of our application. Even though ours will be super simple for now, that's what we'll develop first.

Since we'll be using TypeScript for our application, let's create the interface that will define our Museum object. Follow these steps:

1. Go into src/museums/types.ts and create a type that defines a Museum:

    ```
    export type Museum = {
      id: string,
      name: string,
      description: string,
      location: {
        lat: string,
        lng: string
      }
    }
    ```

    Make sure it is exported as we'll be using this across other files.

    Now that we know the type, we must create some business logic to get a list of museums.

2. Inside `src/museums/types.ts`, create an interface that will define `MuseumController`. It should contain a method that lists all the museums:

```
export interface MuseumController {
    getAll: () => Promise<Museum[]>;
}
```

3. Inside `src/museums/controller.ts`, create a class that will act as the controller. It should contain a function named `getAll`. In the future, this is where the business logic will live, but for now, we can just return an empty array:

```
import type { MuseumController } from "./types.ts";

export class Controller implements MuseumController {
    async getAll() {
        return [];
    }
}
```

We could use this to directly access the database and get certain records. However, since we want to be able to have our business logic isolated and not coupled with other parts of the application, we won't do this.

On top of that, we also want our business logic to be testable in isolation, without depending on a connection to a database or a server. To achieve this, we can't access our data source directly from our controller. Later, we will create an abstraction that will be responsible for getting those records from the database.

For now, we know that we will need to call an external module that will get all the museums for us it give them to our controller – it doesn't matter from where.

Keep in mind the following software design best practice: *"Code to an interface, not an implementation."*

Simply put, this quote means that we should define the module's signature and only then start thinking about its implementation. This vastly helps when it comes to designing clear interfaces.

Going back to our controller, we know that the controller's `getAll` method will, at some point, have to call a module to get the data from a data source.

4. Inside `src/museums/types.ts`, define `MuseumRepository`, the module that will be responsible for getting the museums from a data source:

```
export interface MuseumRepository {
    getAll: () => Promise<Museum[]>
}
```

5. Inside `src/museums/controller.ts`, add an injected class called `museumRepository` to the constructor:

```
import type { MuseumRepository, MuseumController }
  from "./types.ts";

interface ControllerDependencies {
  museumRepository: MuseumRepository
}

export class Controller implements MuseumController {
  museumRepository: MuseumRepository

  constructor({ museumRepository }:
    ControllerDependencies) {
      this.museumRepository = museumRepository
    }

  async getAll() {
    return this.museumRepository.getAll();
  }
}
```

With this, we're stating that whoever instantiates the controller will have to provide a `museumRepository` that implements the `MuseumRepository` interface. By creating this and *lifting the dependencies*, we no longer need to return an empty array from our controller.

Before we write any more logic, let's make sure our code runs and check if it is working. We're just missing one thing.

6. Create a file called `src/index.ts`, import the `MuseumController`, instantiate it, and call the `getAll` method, logging its output. For now, you can inject a dummy repository that just returns an empty array:

```
import { Controller as MuseumController } from
  "./museums/controller.ts";

const museumController = new MuseumController({
  museumRepository: {
    getAll: async () => []
  }
})
console.log(await museumController.getAll())
```

7. Run it to check whether it's working:

```
$ deno run src/index.ts
[]
```

That's it! We just received the empty array that's coming from the dummy repository function!

With the abstraction we have created, our controller is now decoupled from the data source. Its dependencies are injected via a constructor, allowing us to change repositories without changing the controller later.

What we just did is called **dependency inversion** – the **D** in the SOLID principles – and it consists of lifting up part of the dependencies to the function caller. This makes it very easy to test the inside functions independently, as we will see in *Chapter 8*, *Testing – Unit and Integration*, where we'll cover testing.

To transform what we just wrote into a fully functioning app, we need to have a database or something that looks like one. We need something that can store and retrieve a list of museums. Let's create that now.

Developing the data accessing logic

While developing the controller, we noticed that we needed something that would be able to get the data; that is, the repository. This is the module that will abstract all the calls to a data source, and in this case, the data source that stores the museums. It will have a very well-defined set of methods, and whoever wants to access the data should do so through this module.

We already have part of its interface defined inside `src/museums/types.ts`, so let's write a class that implements it. For now, we will not connect it to a real database. We will use an ES6 Map as an in-memory database instead.

Let's get into our repository file and start writing our data accessing logic by following these steps:

1. Open the `src/museums/repository.ts` file and create a `Repository` class.

    It should have a property named `storage` that will be a JavaScript Map. The Map keys should be strings and the values should be objects of the `Museum` type:

    ```
    import type { Museum, MuseumRepository } from
      "./types.ts";


    export class Repository implements MuseumRepository {
      storage = new Map<string, Museum>();
    }
    ```

    We are using TypeScript generics to set the types of our `Map`. Note that we've imported the `Museum` interface from the museum controller, as well as `MuseumRepository`, which is implemented by our class.

    Now that the *database* is "ready", we have to expose certain methods so that people can interact with it. The requirement from the previous section is that we can get all the records from the database. Let's implement that next.

2. Inside the repository class, create a method named `getAll`. It should be responsible for returning all the records in our `storage` Map:

    ```
    export class Repository implements MuseumRepository {
      storage = new Map<string, Museum>();

      async getAll() {
        return [...this.storage.values()];
      }
    }
    ```

We're making it an asynchronous function as we want it to return a promise. This helps us prevent future use cases where we might need to perform asynchronous logic.

Before we proceed, we'll define a pattern that will prove itself to be useful later.

The rule will be that every folder directly inside src should only be accessible from the outside through a single file. This means that whoever wants to import stuff from src/museums should only do so from a single src/museums/index.ts file.

3. Create a src/museums/index.ts file that exports the museum's controller and repository:

```
export { Controller } from "./controller.ts";
export { Repository } from "./repository.ts";


export type { Museum, MuseumController,
  MuseumRepository } from "./types.ts";
```

To remain consistent, we'll need go to all the previous imports that were importing from a file that isn't src/museums/index.ts and change them so that they're only importing things from this file.

4. Update the controller.ts and repository.ts imports to import from the index file:

```
import type { MuseumController, MuseumRepository }
  from "./index.ts";
```

You've probably guessed what we must do next... Do you remember the end of the previous section, where we injected a dummy function into the museum's controller, which was returning an empty array? Let's go back to this and use the logic we just wrote.

5. Go back to src/index.ts, import the Repository class we've just developed and inject it into the MuseumController constructor:

```
import {
  Controller as MuseumController,
  Repository as MuseumRepository,
} from "./museums/index.ts";


const museumRepository = new MuseumRepository();
const museumController = new MuseumController({
  museumRepository })
console.log(await museumController.getAll())
```

Now, let's add a fixture to our "database" so that we can check it if it's actually printing something.

6. Access the storage property in museumRepository and add a fixture to it.

This is currently an anti-pattern as we're directly accessing the module's database, but we'll create a method so that we can add fixtures properly later:

```
const museumRepository = new MuseumRepository();
...
museumRepository.storage.set
    ("1fbdd2a9-1b97-46e0-b450-62819e5772ff", {
    id: "1fbdd2a9-1b97-46e0-b450-62819e5772ff",
    name: "The Louvre",
    description: "The world's largest art museum
        and a historic monument in Paris, France.",
    location: {
        lat: "48.860294",
        lng: "2.33862",
    },
});
console.log(await museumController.getAll())
```

7. Now, let's run our code again:

```
$ deno run src/index.ts
[
    {
        id: "1fbdd2a9-1b97-46e0-b450-62819e5772ff",
        name: "The Louvre",
        description: "The world's largest art
            museum and a historic monument in Paris,
                France.",
        location: { lat: "48.860294", lng: "2.33862" }
    }
]
```

With that, the connection to our database is working, as we can see by the printed fixture.

The abstractions we created in the previous section enabled us to change the data source without changing the controller. This is one of the advantages of the architecture we are using.

Now, if we recall our initial requirement, we can confirm that we are halfway there. Our business logic to satisfy the use case has been created – we're just missing the HTTP part.

Creating the web server

Now that we have our functionality in place, we need to expose it via a web server. Let's use what we've learned from the standard library to create it by following these steps:

1. Create a file named `index.ts` inside the `src/web` folder and add the logic there to create a server. We can copy and paste it from the previous chapter's exercise:

    ```
    import { serve } from
      "https://deno.land/std@0.83.0/http/server.ts";

    const PORT = 8080;
    const server = serve({ port: PORT });

    console.log(`Server running at
      https://localhost:${PORT}`);
    for await (let req of server) {
      req.respond({ body: 'museums api', status: 200 })
    }
    ```

    Since we want our application to be easily configurable, we don't want `port` to be hardcoded here but to be configurable from the outside. We need to export this server creation logic as a function.

2. Wrap the server logic creation inside a function that receives the configuration and `port` as an argument:

    ```
    import { serve } from
      "https://deno.land/std@0.83.0/http/server.ts";

    export async function createServer({
      configuration: {
        port
      }
    ```

```
}) {
  const server = serve({ port });

  console.log(`Server running at
    http://localhost:${port}`);
  for await (let req of server) {
    req.respond({ body: "museums api", status: 200 })
  }
}
```

You've probably noticed that the TypeScript compiler is warning us that we shouldn't use the `port` defining its type.

3. Make this function's parameters an `interface`. This will help us in terms of documentation and will also add type safety and static checks:

```
interface CreateServerDependencies {
  configuration: {
    port: number
  }
}

export async function createServer({
  configuration: {
    port
  }
}: CreateServerDependencies) {
  ...
```

Now that we have configured the web server, we can think of using it for our use case.

4. Go back to `src/index.ts`, import `createServer`, and use it to create a server running on port `8080`:

```
import { createServer } from "./web/index.ts";

...

createServer({
  configuration: {
    port: 8080
```

```
        }
    })
    ...
```

5. Run it and see if it works:

```
$ deno run --allow-net src/index.ts
Server running at http://localhost:8080
[
    {
        id: "1fbdd2a9-1b97-46e0-b450-62819e5772ff",
        name: "The Louvre",
        description: "The world's largest art museum and a
            historic monument in Paris, France.",
        location: { lat: "48.860294", lng: "2.33862" }
    }
]
```

Here, we can see that we have a log stating that the server is running and a log of the result from the previous section.

Now, we can test the web server with `curl` to guarantee it is working:

```
$ curl http://localhost:8080
museums api
```

As we can see, it works – we have some pretty basic logic that still doesn't satisfy our requirements but that spins up a web server. What we'll do next is connect this web server with the logic we wrote previously.

Wiring the web server to the business logic

We're pretty close to finishing what we planned to do at the beginning of this chapter. We currently have a web server and some business logic; it is the connection that is missing.

One quick way to connect both things would be to import the controller directly on `src/web/index.ts` and use it there. Here, the application would have the desired behavior and currently, that doesn't bring any problems.

However, since we are thinking of an app architecture that can grow without many issues, we won't do this. This is because it would make it very hard to test our web logic in isolation, thus compromising one of our principles.

If we import the controller directly from the web server, every time we call the `createServer` function in a testing environment, it will automatically import and call the methods from the `MuseumController` and that's not what we want.

We will use dependency inversion once more to send the controller's methods to the web server. If this still seems too abstract, don't worry – we'll get to the code in a minute.

To make sure we aren't forgetting our initial goal, what we want is for, when a user does a GET request to `/api/museums`, our web server to return a list of museums.

Since we're doing this as an exercise, we will not use a routing library just yet.

We just want to add a basic check to ensure that the URL and method of the request are the ones we want to answer to. If they are, we want to return the list of museums.

Let's go back to the `createServer` function and add our route handler:

```
export async function createServer({
  configuration: {
    port
  }
}: CreateServerDependencies) {
  const server = serve({ port });

  console.log(`Server running at
    http://localhost:${port}`);
  for await (let req of server) {
    if (req.url === "/api/museums" && req.method === "GET")
    {
      req.respond({
        body: JSON.stringify({
          museums: []
        }),
        status: 200
      })
      continue
    }
```

```
      req.respond({ body: "museums api", status: 200 })
  }
}
```

We've added a basic check for the request URL and method and a different response when they match the initial requirement. Let's run the code to see how it behaves:

```
$ deno run --allow-net src/index.ts
Server running at http://localhost:8080
```

Again, test it with `curl`:

```
$ curl http://localhost:8080/api/museums
{"museums":[]}
```

It works – cool!

Now comes the part where we define what we need in order to satisfy this request as an interface.

We ultimately require a function that returns a list of museums to be injected into our server. Let's add that inside the `CreateServerDependencies` interface by following these steps:

1. Back inside `src/web/index.ts`, add `MuseumController` as a dependency to the `createServer` function:

    ```
    import type { MuseumController } from
      "../museums/index.ts";


    interface CreateServerDependencies {
      configuration: {
        port: number
      },
      museum: MuseumController
    }
    ```

Note that we're importing the `MuseumController` type we defined in the museum's module. We're also adding a `museum` object alongside the `configuration` object.

2. Call the `getAll` function from the museum's controller to get a list of all the museums and respond to the request:

```
export async function createServer({
  configuration: {
    port
  },
  museum
}: CreateServerDependencies) {
  const server = serve(`:${port}`);

  console.log(`Server running at
    http://localhost:${port}`);
  for await (let req of server) {
    if (req.url === "/api/museums" &&
      req.method === "GET") {
      req.respond({
        body: JSON.stringify({
          museums: await museum.getAll()
        }),
        status: 200
      })

      continue
    }

    req.respond({ body: 'museums api', status: 200 })
  }
}
```

Now, if we try to run this code, we'll get the following error:

```
$ deno run --allow-net src/index.ts
error: TS2345 [ERROR]: Argument of type '{ configuration:
{ port: number; }; }' is not assignable to parameter of
type 'CreateServerDependencies'.
  Property 'museum' is missing in type '{ configuration:
{ port: number; }; }' but required in type
'CreateServerDependencies'.
createServer({…
```

The TypeScript compiler is right: we have added the museum's controller as a required argument to the createServer function, but we're not sending it when we call createServer. Let's fix that.

3. Go back to src/index.ts, which is where we are calling the createServer function, and send the getAll function from MuseumController. You can also remove the code that directly calls the controller method from the previous section as it is of no use at the moment:

```
import { createServer } from "./web/index.ts";
import {
  Controller as MuseumController,
  Repository as MuseumRepository,
} from "./museums/index.ts";

const museumRepository = new MuseumRepository();
const museumController = new MuseumController({
  museumRepository })

museumRepository.storage.set
  ("1fbdd2a9-1b97-46e0-b450-62819e5772ff", {
  id: "1fbdd2a9-1b97-46e0-b450-62819e5772ff",
  name: "The Louvre",
  description: "The world's largest art museum
    and a historic monument in Paris, France.",
  location: {
    lat: "48.860294",
    lng: "2.33862",
  },
```

```
  });

createServer({
  configuration: { port: 8080 },
  museum: museumController
})
```

4. Run the application again:

    ```
    $ deno run --allow-net src/index.ts
    Server running at http://localhost:8080
    ```

5. Send a request to http://localhost:8080/api/museums; you will get a list
    of museums:

    ```
    $ curl localhost:8080/api/museums
    {"museums":[{"id":"1fbdd2a9-1b97-46e0-b450-
    62819e5772ff","name":"The Louvre","description":"The
    world's largest art museum and a historic monument in
    Paris, France.","location":
    {"lat":"48.860294","lng":"2.33862"}}]}
    ```

And there it is – we're getting the list of museums!

We've just accomplished the goal of this section; that is, to connect our business logic to
the web server.

Note how we've enabled the controller methods to be injected instead of the web
server being the one that directly imports it. This was made possible because we used
dependency inversion. This is something we'll keep doing throughout this book, whenever
we want to decouple and increase the testability of modules and functions.

While doing our mental exercise to test the coupling of our code, when we wanted to use
the current business logic with a different delivery mechanism, such as a CLI, nothing
impeded us. We could still reuse the same controllers and repositories. This means that
we're doing a nice job of using abstractions to decouple our business logic from the
application logic.

Now that we have the baseline of our application architecture and folder structure, and we
also understand the *whys* behind it, we can start looking at the utilities that might help us
build it.

In the next section, we will have a look at the current HTTP frameworks that exist in the Deno community. We won't spend much time doing this, but we want to understand the pros and cons of each one and ultimately choose one to help us with the rest of our journey.

Exploring Deno HTTP frameworks

When you're building an application that's more complex than a simple tutorial, and if you don't want to take a purist approach, you are most likely going to use third-party software.

Obviously, this is not something particular to Deno. Even though there are communities that are keener on using third-party modules than others, all the communities use third-party software.

We could go over the reasons why people do or don't do this, but the more popular reasons are always to do with reliability and time management. This might be because you want to use software that is battle tested instead of building it yourself. Sometimes, it is a mere time management question of not wanting to rewrite something that has already been created.

One important thing we have to say is that we must be extremely cautious in terms of how many of the applications we're building are coupled with third-party software. We don't mean that you should try to reach for the utopia of having everything completely decoupled, especially because that introduces other problems and lots of indirection. What we're saying is that we should be very aware of the cost of bringing a dependency into our code and the trade-offs it introduces.

In the first section of this chapter, we built the foundations for a web application that we'll be adding features to throughout the rest of this book. In its current state, it is still very small, so it doesn't have any dependencies other than the standard library.

In that application, we did a couple of things that we don't believe will scale well, such as defining routes by matching URLs and HTTP methods using plain `if` statements.

As the application grows, it is likely that we will have more advanced needs. These needs can go from dealing with an HTTP request body in different formats to having more complex routing systems, handling headers and cookies, or connecting to a database.

Because we don't believe in reinventing the wheel when it comes to developing applications, we will analyze a few libraries and frameworks that currently exist in the Deno community and are focused on creating web applications.

We will take a general look at the existing solutions and explore their features and approaches.

In the end, we'll choose the one we believe offers the best trade-off for our use case.

What alternatives exist?

At the time of writing, there are a few third-party packages that offer a great amount of functionality to create web applications and APIs. Some of them are heavily inspired by very popular Node.js packages, such as Express.JS, Koa, or hapi.js, while others are inspired by other frameworks outside of JavaScript, such as Laravel, Flask, and so on.

We'll be exploring four of them that are quite popular and well-maintained at the time of writing. Keep in mind since as Deno and the mentioned packages are evolving quickly, this might change over time.

> **Important note**
> There is a great article by Craig Morten that does a very thorough analysis and exploration of the available libraries. I heavily recommend this article if you want to find out more (`https://dev.to/craigmorten/what-is-the-best-deno-web-framework-2k69`).

We will try to be diverse when it comes to the packages we're going to explore. There are some that provide more abstractions and structure than others, and some that provide not much more than mere utility functions and composable functionality.

The packages we'll explore are as follows:

- Drash
- Servest
- Oak
- Alosaur

Let's look at each one separately.

Drash

Drash (`https://github.com/drashland/deno-drash`) aims to be different from the existing Deno and Node.js frameworks. This motivation is explicitly mentioned by its maintainer, Edward Bebbington, on a blog post where he compares Drash to other alternatives and explains the motivations behind its creation (`https://dev.to/drash_land/what-makes-drash-different-idd`).

These motivations are great, and the inspiration of very popular software tools such as Laravel, Flask, and Tonic justifies most of these decisions. Some of the similarities are also noticeable the moment you look at Drash's code.

It really offers a different approach compared to libraries such as Express.js or Koa, as the documentation states:

> *"Where Deno is different than Node.js, Drash aims to be different than Express or Koa, utilizing resources and a full class-based system."*

The main differences are that Drash doesn't want to provide an application object where developers can then register their endpoints, like some popular Node.js frameworks do. It sees endpoints as resources that are defined within a class, similar to the following:

```
import { Drash } from
  "https://deno.land/x/drash@v1.2.2/mod.ts";


class HomeResource extends Drash.Http.Resource {
  static paths = ["/"];
  public GET() {
    this.response.body = "Hello World!";
    return this.response;
  }
}
```

These resources are then plugged into Drash's application later:

```
const server = new Drash.Http.Server({
  response_output: "text/html",
  resources: [HomeResource]
});


server.run({
  hostname: "localhost",
```

```
    port: 1447
});
```

Here, we can directly state it is, in fact, different from the other frameworks we've mentioned. These differences are deliberate and plan to please developers that are fans of this approach and the problems it solves for other frameworks. Those use cases are very well-explained in Drash's documentation.

Drash's resource-based approach is definitely something to keep an eye on. Its inspiration from very mature pieces of software such as Flask and Tonic definitely brings something to the table and proposes a solution that helps solve some of the common problems unopinionated tools have. The documentation is complete and easy to understand, which makes it a great asset to have when it comes to choosing a tool for building your application with.

Servest

Servest (`https://servestjs.org/`) calls itself a *"progressive HTTP server for Deno."*

One of the reasons it was created was because its author wanted to make some APIs from the standard library's HTTP module easier to use and experiment with new features. The latter is something that is really hard to do on a standard library that needs stability.

Servest directly focuses on this comparison with the standard library's HTTP module. One of its main objectives, which is directly stated on the project's home page, is making it easy to migrate from the standard library's HTTP module to Servest. This summarizes Servest's vision well.

API-wise, Servest is very similar to what we're used to from Express.js and Koa. It provides an application object where routes can be registered. You can also recognize obvious influences from what is provided by the standard library module, as we can see in the following code snippet:

```
import { createApp } from
    "https://servestjs.org/@v1.1.4/mod.ts";
const app = createApp();
app.handle("/", async (req) => {
  await req.respond({
    status: 200,
    headers: new Headers({
      "content-type": "text/plain",
    }),
```

```
    body: "Hello, Servest!",
  });
});
app.listen({ port: 8899 });
```

We can recognize the application object from well-known Node.js libraries and the request object from the standard library, among other things.

On top of this functionality, Servest also provides common features, such as support for directly rendering JSX pages, serving static files, and authentication. The documentation is also clear and full of examples.

Servest tries to leverage knowledge and familiarity from Node.js users while using the benefits provided by Deno in a promising mix. Its progressive nature brings very nice features to the table, with the promise to make developers more productive than when they're using the standard library HTTP package.

Oak

Oak (`https://oakserver.github.io/oak/`) is currently the most popular Deno library when it comes to creating web applications. Its name derives from a play on words of Koa, a very popular Node.js middleware framework and Oak's main inspiration.

Due to its heavy inspirations, it is of no surprise that its APIs resemble Koa by using asynchronous functions and a context object. Oak also includes a router, also inspired by `@koa/router`.

If you know Koa, the following code might look very familiar to you:

```
import { Application } from
  "https://deno.land/x/oak/mod.ts";

const app = new Application();

app.use((ctx) => {
  ctx.response.body = "Hello world!";
});

await app.listen("127.0.0.1:8000");
```

For those of you who are not familiar with Koa, we'll explain it in brief, since understanding it will help you understand Oak.

Koa provides a minimal and unopinionated approach by using modern JavaScript features. One of the initial reasons Koa was created (by the same team who created Express.js) was that its creator wanted to create a framework that would leverage modern JavaScript features, as opposed to Express, which was created at the beginning of Node.js' lifetime.

The team wanted to use new features such as promises and async/await, and then solve challenges that developers faced with Express.JS. Most of these challenges were related to error handling, dealing with callbacks, and the lack of clarity of some APIs.

Oak's popularity doesn't come out of nowhere, and its current distance from the alternatives in terms of GitHub stars reflects that. By themselves, GitHub stars don't mean much, but together with opened and closed issues, the number of releases, and so on, we can see why people are trusting it. Of course, this familiarity plays a big role in terms of this package's popularity.

In its current state, Oak is a solid (in terms of Deno's community standards) way to build web applications as it provides a very clear and direct set of features.

Alosaur

Alosaur (`https://github.com/alosaur/alosaur`) is a Deno web application framework based on decorators and classes. It is similar to Drash in a way, even though the final approaches are quite different.

Among its main features, Alosaur provides things such as template rendering, dependency injection, and OpenAPI support. These features have been added on top of what is a standard for all the alternatives we presented here, such as middleware support and routing.

This framework's approach is to define controllers by using classes and define their behavior using decorators, as shown in the following code:

```
import { Controller, Get, Area, App } from
  'https://deno.land/x/alosaur@v0.21.1/mod.ts';


@Controller() // or specific path @Controller("/home")
export class HomeController {
    @Get() // or specific path @Get("/hello")
    text() {
```

```
        return 'Hello world';
    }
}

// Declare module
@Area({
    controllers: [HomeController],
})
export class HomeArea {}

// Create alosaur application
const app = new App({
    areas: [HomeArea],
});

app.listen();
```

Here, we can see that the application's instantiation has similarities with Drash. It also uses TypeScript decorators to declare a framework's behavior.

Alosaur takes a different approach compared to most of the aforementioned libraries, mainly because it doesn't try to be minimal. Instead, it provides a set of features that prove to be useful when it comes to building certain types of applications.

We decided to have a look at it not only because it does what it is supposed to do, but also because it has some features that are not so common in the Node.js and Deno world. This includes things such as dependency injection and OpenAPI support, which are not offered by any other of the presented solutions. At the same time, it keeps features such as template rendering, which is something that might be familiar to you from Express.JS but not so familiar in more modern frameworks.

The final solution is very promising and complete in the sense of the functionalities it provides. It is definitely something to keep an eye on so that you can see how it evolves.

The verdict

After looking at all the presented solutions and recognizing that all of them have their merits, we've decided to go with Oak for the rest of this book.

This doesn't mean that this book will focus on Oak. It will not, as it will only handle HTTP and routing. Oak's minimal approach will fit very nicely with what we will be doing next, helping us to incrementally create features without it getting in the way. The fact that it is also one of the most stable, maintained, and popular options in the Deno community has an obvious influence on our decision.

Be aware that this decision doesn't mean that what we will learn in the next few chapters can't be done in any of the alternatives. In fact, because of the way we will organize our code and architecture, we believe that it would be easy to keep up with most of the things we are going to do using a different framework.

Throughout the rest of this book, we will use other third-party modules that can help us build the functionalities we've proposed. The reason we've decided to have a deeper look at the libraries that deal with HTTP is because this is the fundamental delivery mechanism for the application we'll be developing.

Summary

In this chapter, we finally started building an application that leverages our knowledge of Deno. We started by considering the main goals we will have when we build an application and define its architecture. These goals will set the tone for most of our conversations regarding architecture and structure throughout this book as we'll keep going back to them, ensuring that we're in line with them.

We started by creating our folder structure and trying to achieve our very first application goal: have an HTTP endpoint that lists museums. We did this by building the simple business logic first and progressed while needs such as separation of concerns and isolation and responsibilities arose. These needs derived our architecture, proving why the layers and abstractions we created are useful and demonstrating what they add.

By having the responsibilities and the module's interfaces well-defined, we understood that we could temporarily build our application by using an in-memory database, which we did. It was possible to build the application to fit this chapter's requirement with that, and layer separation allows us to come back later and change it to a proper persistency layer without any issues. After defining the business and data accessing logic, we created a web server with the standard library that worked as a delivery mechanism. After creating a very rudimentary routing system, we plugged in the business logic we built previously and satisfied the main requirement for this chapter: having an application that returns a list of museums.

We did all of this without creating a direct coupling between the business logic, the data fetching, and the delivery layers. This is something we believe will be very useful as we start adding complexity, extending our application, and adding tests to it.

This chapter wraps up by having a look at the HTTP frameworks and libraries that currently exist in the Deno community and understanding their differences and approaches. Some of them use approaches that are familiar to Node.js users, while others deeply use TypeScript and its features to create a more structured web application. By looking at the four currently available solutions, we learned what is being developed in the community and the directions they might go in.

We ended up choosing Oak, a very minimal and reasonably mature solution, to help us solve the routing and HTTP challenges we'll come across in the rest of this book.

In the next chapter, we'll start adding Oak to our code base, together with useful features such as authentication and authorization, using concepts such as middleware, and growing our application so that it achieves the goals we set ourselves up to complete.

Let's go!

5
Adding Users and Migrating to Oak

At this point, we have laid the foundations for a web application with a structure that will enable us to add more functionalities as we proceed. As you might have guessed by the name of this chapter, we'll start this chapter by adding the middleware framework of our choice to the current web application, Oak.

Together with Oak, and since our application is starting to have more third-party dependencies, we'll use what we've learned in previous chapters to create a lock file and perform integrity checking when installing dependencies. This way, we can guarantee that our applications will run smoothly without dependency problems.

As we get into this chapter, we'll start understanding how to simplify our code using Oak's features. We'll make our routing logic more extendable but also more scalable. Our first solution was to use `if` statements together with the standard library to create a DIY routing solution, which we'll refactor here.

Once we've done this, we'll end up with much cleaner code and be able to use Oak's features, such as automatic content-type definition, handling unallowed methods, and route prefixing.

Then, we'll add a feature that is crucial in pretty much every application: users. We'll create a module alongside museums to handle everything that is user-related. In this new module, we'll develop the business logic to create users, as well as the code to create new users in the database, by using common practices such as hashes and salts.

While implementing these features, we will get to know other modules provided by Deno, such as the hashing features of the standard library or the crypto APIs that are included in the runtime.

Adding this new module and having it interact with the rest of the application will be a nice way to test the application architecture. By doing this, we'll understand how it scales while keeping everything that is related to a context in a single place.

The following topics will be covered in this chapter:

- Managing dependencies and lock files
- Writing a web server with Oak
- Adding users to the application
- Let's get started!

Technical requirements

This chapter will build on top of the code we developed in the previous chapter. All the codes files for this chapter are available in this book's GitHub repository at `https://github.com/PacktPublishing/Deno-Web-Development/tree/master/Chapter05/sections`.

Managing dependencies and lock files

In *Chapter 2, The Toolchain*, we learned how Deno enables us to do dependency management. In this section, we'll use it in a more practical context. We'll start by removing all the scattered imports with URLs from our code and move them into a central dependency file. After this, we'll create a lock file that makes sure our still young application runs smoothly anywhere it is installed. We'll finish by learning how can we install the project's dependencies based on a lock file.

Using a centralized dependency file

In the previous chapter, you probably noticed that we were using direct URLs to dependencies directly in our code. Even though this is possible, this was something we discouraged a few chapters ago. It worked for us in that first phase, but as the application starts growing, we'll have to manage our dependencies properly. We want to avoid struggles with conflicting dependency versions, typos in the URLs, and having dependencies scattered across files. To solve this, we must do the following:

1. Create a `deps.ts` file at the root of the `src` folder.

    This file can have whatever name you prefer. We're currently calling it `deps.ts` as it is what is mentioned in Deno's documentation, and it's the naming convention many modules use.

2. Move all the external dependencies from our code to `deps.ts`.

    Currently, the only dependency we have is the HTTP module from the standard library, which can be found in the `src/web/index.ts` file.

3. Move the import into the `deps.ts` file and change `import` to `export`:

    ```
    export { serve } from
        "https://deno.land/std@0.83.0/http/server.ts"
    ```

4. Notice how the fixed version is on the URL:

    ```
    export { serve } from
        "https://deno.land/std@0.83.0/http/server.ts"
    ```

    This is how versioning works in Deno, as we learned in *Chapter 2, The Toolchain*.

    We now need to change the dependent files so that they import directly from the `deps.ts` file instead of being directly imported from the URL.

5. In `src/web/index.ts`, import the `serve` method from `deps.ts`:

    ```
    import { serve } from "../deps.ts";
    ```

By having a centralized dependency file, we also have an easy way to make sure we have all our dependencies locally downloaded without having to run any code. With this, we now have a single file where we can run the `deno cache` command (mentioned in *Chapter 2, The Toolchain*).

Creating a lock file

Having centralized our dependencies, we need to guarantee that whoever installs the project is getting the same versions of the dependencies we did. This is the only way we can guarantee that the code will run in the same way. We'll do this by using a lock file. We learned how to do this in *Chapter 2, The Toolchain*; here, we'll apply it to our application.

Let's run the `cache` command with the `lock` and `lock-write` flags, plus a path to the lock file and a path to the centralized dependencies file, `deps.ts`:

```
$ deno cache --lock=lock.json --lock-write src/deps.ts
```

A `lock.json` file should be generated in the current folder. If you open it, it should contain a key-value pair of the URL, as well as the hash that's used for performing integrity checks.

This lock file should then be added to version control. Later, if a coworker wants to install this same project, they just have to run the same command without the `--lock-write` flag:

```
$ deno cache --lock=lock.json src/deps.ts
```

With that, the dependencies in `src/deps.ts` (this should be all of them) will be installed and have their integrity checked according to the `lock.json` file.

Now, every time we install a new dependency in the project, we must run the `deno cache` command with the `lock` and `lock-write` flags, to make sure the lock file is updated.

That's pretty much it for this section!

In this section, we learned a simple but very important step in making sure an application runs smoothly. This helps us avoid future hairy problems such as dependency conflicts and mismatches in behavior across versions. We're also guaranteeing resource integrity, something that is even more important in Deno, since its dependencies are hosted in a URL instead of a registry.

In the next section, we'll start *refactoring* our application from the standard library HTTP server into Oak, which will result in our web code being simplified.

Writing a web server with Oak

At the end of the previous chapter, we looked at different web libraries. After a brief analysis, we ended up choosing Oak. In this section, we'll rewrite part of our web application so that we can use it instead of the HTTP module from the standard library.

Let's open `src/web/index.ts` and start tackling it step by step.

Following Oak's documentation (`https://deno.land/x/oak@v6.3.1`), the only thing we'll need to do is instantiate the `Application` object, define a middleware, and call the `listen` method. Let's do it:

1. Add Oak's import to the `deps.ts` file:

    ```
    export { Application } from
      "https://deno.land/x/oak@v6.3.1/mod.ts"
    ```

    If you are using VSCode, then you've probably noticed that there is warning saying that it couldn't find this version of the dependency locally.

2. Let's run the commands from the previous section to download it and add it to the lock file.

    Do not forget to do this every time we add a dependency so that we have better autocompletion and our lock file is always updated:

    ```
    $ deno cache --lock=lock.json --reload --lock-write src/
    deps.ts
    Download https://deno.land/std@0.83.0/http/server.ts
    Download https://deno.land/x/oak@v6.3.1/mod.ts
    Download https://deno.land/std@0.83.0/encoding/utf8.ts
    ...
    ```

    With all the necessary dependencies downloaded, let's start using them in our code.

3. Delete all the code from the `createServer` function in `src/web/index.ts`.

4. Inside `src/web/index.ts`, import the `Application` class and instantiate it. Create a very simple piece of middleware (as mentioned in the documentation) and call the `listen` method:

    ```
    import { Application } from "../deps.ts";
    ...
    export async function createServer({
      configuration: {
        port
      },
      museum
    }: CreateServerDependencies) {
      const app = new Application ();
    ```

```
    app.use((ctx) => {
      ctx.response.body = "Hello World!";
    });


    await app.listen({ port });
  }
```

Keep in mind that, while removing the old code, we also removed `console.log` and thus it will not print anything just yet. Let's run it and verify that it has no problems:

```
$ deno run --allow-net src/index.ts
```

Now, if we access `http://localhost:8080`, we'll see the "Hello World!" response there.

Now, you might be wondering what the `use` method from Oak's application is. Well, we'll be using this method to define middleware. For now, we just want it to modify the response and add a message to its body. In the next chapter, we'll learn about middleware functions in more depth.

Remember when we've removed `console.log` and that we don't get any feedback if the application is running? We'll learn how to do this while we learn about how to add event listeners to an Oak application.

Adding event listeners to an Oak application

So far, we've managed to get the application to run, but at the moment, we don't have any message to acknowledge this. We'll use this as an excuse to learn about event listeners in Oak.

Oak applications dispatch two different types of events. One of them is `listen`, while the other is `the` `listen` event is the one we'll use to log to the console when an application is running. The other, `error`, is the one we'll use to log to the console when an error occurs.

First, let's add the event listener for the `listen` event, before the `app.listen` statement:

```
app.addEventListener("listen", e => {
  console.log(`Application running at
    http://${e.hostname || 'localhost'}:${port}`)
```

```
})
…
await app.listen({ port });
```

Note that we're not only printing a message to the console but also printing `hostname` from the event and sending it a default value, in case it is undefined.

For safety and to guarantee that we catch any unexpected errors, let's also add an error event listener. This error event will be triggered in case an error that hasn't been handled occurs in our application:

```
app.addEventListener("error", e => {
    console.log('An error occurred', e.message);
})
```

These handlers, especially the `error` one, will help us a lot when we're developing and want to gather quick feedback about what's happening. Later, when closer to the production stage, we'll add proper logging middleware.

Now, you might be thinking that we're still missing the functionality we had when we started this chapter, and you'd be right: we've removed the endpoint that listed all the museums from our application.

Let's add it again and learn how can we create routes in an Oak application.

Handling routes in an Oak application

Oak provides another object, alongside the `Application` class, that allows us to define routes – the `Router` class. We'll use this to reimplement the route we had before, which listed all the museums in the application.

Let's create it by sending the prefix property to the constructor. Doing this means that all the routes defined there will be prefixed with that path:

```
import { Application, Router } from "../deps.ts";
…
const apiRouter = new Router ({ prefix: "/api" })
```

Now, let's get back our functionality, which returns the list of museums via a GET request to /api/museums:

```
apiRouter.get ("/museums", async (ctx) => {
    ctx.response.body = {
```

```
      museums: await museum.getAll()
  }
});
```

A few things are happening here.

Here, we're defining a route using Oak's router API by sending a URL and a handler function. Our handler is then called with a context (`ctx`) object. All of this is explained in detail in Oak's documentation (`https://doc.deno.land/https/deno.land/x/oak@v6.3.1/mod.ts#Router`), but I'll leave you with a short resume.

In Oak, everything that a handler can do is done through the context object. The request that's made is available in the `ctx.request` property, while the response for the current request is available in `ctx.response`. Things such as headers, cookies, parameters, the body, and so on are available in those objects. Some properties, such as `ctx.response.body`, are writable.

> **Tip**
> You can get a better overview of Oak's functionality by looking at Deno's documentation website: `https://doc.deno.land/https/deno.land/x/oak@v6.3.1/mod.ts`.

In this case, we're using the response body property to set its content. When Oak can infer the response's type (which is JSON here), it automatically adds the correct `Content-Type` header to the response.

We'll be learning more about Oak and its features throughout this book. The next step is to connect our recently created router.

Connecting the router to the application

Now that our router is defined, we need to register it on the application so that it can start handling requests.

To do that, we'll use a method of the application instance we've used previously – the `use` method.

In Oak, once a `Router` has been defined (and its registered), it provides two methods that return middleware functions. These functions can then be used to register routes on the application. They are as follows:

- `routes`: Registers the registered route handlers in the application.

- `allowedMethods`: Registers automatic handlers for the API calls of methods that are not defined in the router, returning a `405 - Not allowed` response.

We'll use both of them to register our router in the main application, as follows:

```
const apiRouter = new Router({ prefix: "/api" })

apiRouter.get("/museums", async (ctx) => {
  ctx.response.body = {
    museums: await museum.getAll()
  }
});

app.use(apiRouter.routes());
app.use(apiRouter.allowedMethods());

app.use((ctx) => {
  ctx.response.body = "Hello World!";
});
```

And with this, our router registers its handlers in the application, and they are ready to start handling requests.

Keep in mind that we have to register these before the Hello World middleware we defined earlier. If we don't do this, the Hello World handler will respond to all the requests before they reach our router, and thus it will not work.

Now, we can run our application by running the following command:

```
$ deno run --allow-net src/index.ts
Application running at http://localhost:8080
```

Then, we can perform a `curl` to the URL:

```
$ curl http://localhost:8080/api/museums
{"museums":[{"id":"1fbdd2a9-1b97-46e0-b450-
62819e5772ff","name":"The Louvre","description":"The world's
largest art museum and a historic monument in Paris,
France.","location":{"lat":"48.860294","lng":"2.33862"}}]}
```

As we can see, everything is working as expected! We've managed to migrate our application to Oak.

By doing this, we've vastly improved the readability of our code. We also used Oak to handle stuff we didn't wanted to deal with, and we managed to focus on our application.

In the next section, we will add the concept of users to the application. More routes will be created, as well as a whole new module and some business logic to handle users.

> **Tip**
>
> The code from this chapter is available, separated by sections, at `https://github.com/PacktPublishing/Deno-Web-Development/tree/master/Chapter05/sections`.

Now, let's add some users to the application!

Adding users to the application

We currently have the first endpoint running and listing all the museums in the application, but we're still far from meeting the final requirements.

We want to add users so that it is possible to register, log in, and interact with the application with an identity.

We'll start by creating the object that will define the user, and then proceed into the business logic to create and store it. After this, we'll create endpoints that will allow us to interact with the application via HTTP, thus allowing users to register.

Creating the user module

We currently have what we can call a single "module" in the application: the `museums` module. Everything that is related to museums is there, from controllers to repositories, object definitions, and so on. This module has one single interface, which is its `index.ts` file.

We did this so that we have the freedom of working inside the module while maintaining its external API so that it's always stable. This gives us a nice degree of decoupling between modules. To make sure that the pieces inside a module are reasonably decoupled, we must also inject their dependencies via a constructor, allowing us to easily swap pieces and test them in isolation (as you'll see in *Chapter 8, Testing – Unit and Integration*).

Following those guidelines, we'll keep using this "modules" system and create one for our users by following these steps:

1. Create a folder named `src/users` and put the `index.ts` file inside it.

2. Create the `src/users/types.ts` file. This is where we'll define the `User` type:

    ```
    export type User = {
      username: string,
      hash: string,
      salt: string,
      createdAt: Date
    }
    ```

    Our user object will be very simple: it will have a `username`, a `createdAt` date, and then two properties: `hash` and `salt`. We'll use these to safeguard the user's password when we store it.

3. Create the user controller in `src/users/controller.ts` with a `register` method. It should receive a username and a password, and then create a user in the database:

    ```
    type RegisterPayload =
      { username: string, password: string };


    export class Controller {
      public async register(payload: RegisterPayload) {
        // Logic to register users
      }
    }
    ```

4. Define `RegisterPayload` in `src/users/types.ts` and export it in `src/users/index.ts`, removing it from `src/users/controller.ts`

Inside `src/users/types.ts`, add the following:

```
// src/users/types
export type RegisterPayload =
  { username: string; password: string };
```

Inside `src/users/index.ts`, add the following:

```
export type {
  RegisterPayload,
} from "./types.ts";
```

Let's stop here for now and think about the register logic.

To create a user, we must check if that user exists in the database. If they don't, we'll create them with the username and password that was entered, and then return an object that doesn't contain sensitive data.

In the previous chapter, we used the repository pattern every time we wanted to interact with a data source. The repository kept all the *data accessing* logic (`src/museums/repository.ts`).

Here, we're going to do the same. We've already noticed that our controller needs to call two methods in `UserRepository`: one to check if a user exists and another to create the user. That's the interface we're going to be defining next.

5. Go to `src/users/types.ts` and define the interface for `UserRepository`:

```
export type CreateUser =
  Pick<User, "username" | "hash" | "salt">;

...

export interface UserRepository {
  create: (user: CreateUser) => Promise<User>
  exists: (username: string) => Promise<boolean>
}
```

Note how we created a `CreateUser` type that contains all the properties of `User` except for `createdAt`, which should be added by the repository.

With the `UserRepository` interface defined, we can now move on to our user's controller and make sure it receives an instance of the repository in the constructor.

6. In `src/users/controller.ts`, create a `constructor` that receives the user repository as an injected parameter and sets the class property with the same name:

```
import type { UserRepository } from "./types.ts";

type RegisterPayload =
  { username: string, password: string };
...
interface ControllerDependencies {
  userRepository: UserRepository;
}

export class Controller {
  userRepository: UserRepository;

  constructor({ userRepository }:
    ControllerDependencies) {
    this.userRepository = userRepository;
  }
  public async register(payload: RegisterPayload) {
    // Logic to register users
  }
}
```

Now that we have access to `userRepository`, we can start writing the logic for the `register` method.

7. Write the logic for the `register` method, check if the user exists, and create them if not:

```
async register(payload: RegisterPayload) {
  if (await
   this.userRepository.exists(payload.username)) {
     return Promise.reject("Username already exists");
  }

  const createdUser = await
    this.userRepository.create(
    { username: payload.username,
```

```
        hash: "random-hash", salt: "random-salt" }
  )

    return createdUser;
  }
```

Note how we're sending a direct string as hash and salt properties to the `create` method of `userRepository` to make sure it follows the `CreateUser` type we defined previously. These will have to be automatically generated, but don't worry about that for now.

And with this, we've pretty much finished looking at what will happen whenever someone tries to register with our application.

We're still missing one thing, though. As you may have noticed, we're returning the `User` object directly from the repository, which might contain sensitive information, namely the `hash` and `salt` properties.

8. Create a type called `UserDto` in `src/users/types.ts` that defines the format of the `User` object without sensitive data:

```
export type User = {
  username: string,
  hash: string,
  salt: string,
  createdAt: Date
}

export type UserDto = Pick<User, "createdAt" |
  "username">
```

Note that we're using TypeScript's `Pick` to choose two properties from the `User` object; that is, `createdAt` and `username`.

With `UserDto` (https://en.wikipedia.org/wiki/Data_transfer_object) defined, we can now make sure our register is returning it.

9. Create a file called `src/users/adapter.ts` with a function called `userToUserDto` inside it that converts a user into a `UserDto`:

```
import type { User, UserDto } from "./types.ts";

export const userToUserDto = (user: User): UserDto => {
  return {
    username: user.username,
    createdAt: user.createdAt
  }
}
```

10. Use the recently created function in the register method to make sure we're returning a `UserDto`:

```
import { userToUserDto } from "./adapter.ts";
...
public async register(payload: RegisterPayload) {
  ...
  const createdUser = await
    this.userRepository.create(
    payload.username,
    payload.password
  );

  return userToUserDto(createdUser);
}
```

With that, the `register` method is complete!

We're currently sending the hash and salt as two plain strings that don't mean anything.

You might be wondering why we don't send the password directly. This is because we want to make sure we're not storing passwords in plain text in any database.

To make sure we're following the best practices, we will use hashing and salting to store the users' password in the database. While doing that, we also want to learn about a couple more Deno APIs. That's what we'll do in the next section.

Storing a user in the database

Even though we're using an in-memory database, we've decided that we won't store the passwords in plain text. Instead, we'll use a common method to store passwords called hashing and salting. If you are not familiar with it, auth0 has a great article on it that I definitely recommend (`https://auth0.com/blog/adding-salt-to-hashing-a-better-way-to-store-passwords/`).

The pattern itself is not complicated, and you can learn it just by following the code.

So, what we will do is store our password hashed. We won't be storing the exact hashed password the user entered, but the password plus a randomly generated string, called a salt. This salt will then be stored alongside the password so that it can be used later. After this, we will never need to decode the password again.

With the salt, any time we want to check if a password is correct, we just have to add the salt to whatever password the user entered, hash it, and verify that the output matches what is stored in the database.

If this still seems strange to you, I can guarantee it becomes much simpler when you look at the code. Let's implement these functions by following these steps:

1. Create a utils file called `src/users/util.ts` with a `hashWithSalt` function inside it that hashes a string with the provided salt:

> **Tip**
>
> The standard library provides a set of hashing methods at `https://deno.land/std@0.83.0/hash`.

```
import { createHash } from
  "https://deno.land/std@0.83.0/hash/mod.ts";

export const hashWithSalt =
  (password: string, salt: string) => {
    const hash = createHash("sha512")
      .update(`${password}${salt}`)
        .toString();

    return hash;
  };
```

It should be clear by now that this function will return a string that is the hash value of the provided string, plus a salt.

It's also considered a best practice (as mentioned in the article mentioned previously) to use different salts for different passwords. By generating a different salt for each password, we ensure all the passwords are still safe if one password's salt is leaked.

Let's proceed by creating a function that will generate a salt.

2. Create a generateSalt function using the crypto API (https://doc.deno.land/builtin/stable#crypto) to get random values and generate a salt string from there:

> **Tip**
>
> The standard library provides a set of methods we can use to encode values from and to different formats that you may find useful: https://doc.deno.land/https/deno.land/std@0.83.0/encoding/hex.ts.

```
import { encodeToString } from
  "https://deno.land/std@0.83.0/encoding/hex.ts"

...

export const generateSalt = () => {
  const arr = new Uint8Array(64);
  crypto.getRandomValues(arr)

  return encodeToString(arr);
}
```

And that's all we need to generate hashed passwords for our application.

Now, we can start using the utility functions we just created in our controller. Let's create a method so that we can hash our password there.

3. Create a private method inside UserController called getHashedUser that receives a username and password and returns a user, along with their hash and salt:

```
import { generateSalt, hashWithSalt } from
  "./util.ts";

...

export class Controller implements UserController {
```

```
...
    private async getHashedUser
      (username: string, password: string) {
    const salt = generateSalt();
    const user = {
      username,
      hash: hashWithSalt(password, salt),
      salt
    }

    return user;
  }
...
```

4. Use the recently created getHashedUser method in the register method:

```
public async register(payload: RegisterPayload) {
  if (await
    this.userRepository.exists(payload.username)) {
    return Promise.reject("Username already exists");
  }

  const createdUser = await
    this.userRepository.create(
    await this.getHashedUser
      (payload.username, payload.password)
  );

  return userToDto(createdUser);
}
```

And we're done! With that, we've made sure that we aren't storing any plain text passwords. In the path, we learned about the crypto APIs available in Deno.

We did all of this implementation while using on the UserRepository interface that we defined previously. However, currently, we don't have a class that implements it, so let's create one.

Creating the user repository

In the previous section, we created the interface that defined `UserRepository`, so next, we're going to create a class that implements it. Let's get started:

1. Create a file called `src/users/repository.ts` with an exported `Repository` class inside it:

    ```
    import type { CreateUser, User, UserRepository } from
        "./types.ts";

    export class Repository implements UserRepository {
      async create(user: CreateUser) {
      }

      async exists(username: string) {
      }
    }
    ```

    The interface guarantees that these two public methods need to exist.

    Now, we need a way to store the users. For the purpose of this chapter, we'll use an in-memory database again, very similar to what we did with our museums.

2. Create a property inside the `src/users/repository.ts` class called `storage`. It should be a JavaScript Map, and it will work as the users' database:

    ```
    import { User, UserRepository } from "./types.ts";

    export class Repository implements UserRepository {
      private storage = new Map<User["username"], User>();
      ...
    ```

    With the database in place, we can now implement the logic for the two methods.

3. Get the user from the database in the `exists` method, returning `true` if it is there and `false` if not:

    ```
    async exists(username: string) {
      return Boolean(this.storage.get(username));
    }
    ```

The Map#get function returns undefined if it can't get the record, so we're converting it into a Boolean to make sure it always returns true or false.

The exists method is quite simple; it just needs to check whether the user is present in the database and a boolean is returned accordingly.

To create a user, we need do perform one or two steps more than that. More than just creating it, we'll have to ensure it also adds a createdAt date to the user that's sent by whoever is calling this function.

Now, let's go back and complete our main task: creating a user in the database.

4. Open the src/users/repository.ts file and implement the create method, creating a user object in the proper format.

    Remember to add createdDate to the user object that was sent to the function:

    ```
    async create(user: CreateUser) {
      const userWithCreatedAt =
        { ...user, createdAt: new Date() }
      this.storage.set
        (user.username, { ...userWithCreatedAt });
      return userWithCreatedAt;
    }
    ```

And with that, our repository is complete!

It fully implements what we previously defined in the UserRepository interface and is ready to be used.

The next step is to wire all these pieces together. We've already created the User controller and the User repository, but they're still not being used anywhere.

Before we proceed, we need to expose these objects from the user module to the outside world. We'll follow the rule we defined previously; that is, the modules interface will always be an index.ts file at its root.

5. Open src/users/index.ts and export the Controller, the Repository classes, and their respective types from the module:

    ```
    export { Repository } from './repository.ts';
    export { Controller } from './controller.ts';

    export type {
      CreateUser,
    ```

```
    RegisterPayload,
    User,
    UserController,
    UserRepository,
} from "./types.ts";
```

We can now make sure that every file in the user module is importing types directly from this file (`src/users/index.ts`) instead of going directly to other files.

Now, any module that wants to import stuff from the user module must do so through the `index.ts` file. Now, we can start to think about how the user will interact with the business logic we just wrote. Since we're building an API, we'll learn how to expose it via HTTP in the next section.

Creating the register endpoint

With the business logic and data accessing logic ready, the only thing missing is the endpoint that the user can call to register itself.

For the register request, we'll implement a `POST /api/users/register` expecting a JSON object with a property named `user` containing two properties, `username` and `password`.

The first thing we'll have to do is declare that our `createServer` function in `src/web/index.ts` will depend on the `UserController` interface to be injected. Let's get started:

1. In `src/users/types.ts`, create the `UserController` interface. Make sure it is also exported in `src/users/index.ts`:

    ```
    export type RegisterPayload =
      { username: string, password: string };
    export interface UserController {
      register: (payload: RegisterPayload) =>
    Promise<UserDto>
    }
    ```

    Remember that we also moved `RegisterPayload` from `src/users/controller.ts` previously.

2. Now, just to keep things tidy, go to `src/users/controller.ts` and make sure that the class implements `UserController`:

```
import { RegisterPayload, UserController,
  UserRepository } from "./types.ts";


export class Controller implements UserController
```

3. Back inside `src/web/index.ts`, add `UserController` to the `createServer` dependencies:

```
import { UserController } from "../users/index.ts";


interface CreateServerDependencies {
  configuration: {
    port: number
  },
  museum: MuseumController,
  user: UserController
}


export async function createServer({
  configuration: {
    port
  },
  museum,
  user
}: CreateServerDependencies) {
  ...
```

We're now ready to create our register handler.

4. Create a handler that responds to a `POST` request in `/api/users/register` and creates a user using the injected controller's `register` method:

```
apiRouter.post("/users/register", async (ctx) => {
  const { username, password } = await
    ctx.request.body({ type: 'json' }).value;

  if (!username || !password) {
```

```
        ctx.response.status = 400;

        return;
    }

    try {
        const createdUser = await user.register({
            username, password });
        ctx.response.status = 201;
        ctx.response.body = { user: createdUser };
    } catch (e) {
        ctx.response.status = 400;
        ctx.response.body = { message: e.message };
    }
});
```

A few things are happening here. Let's break this down.

First, we're using the `post` method to define a route that accepts a POST request. Then, we're using the `body` method from the request (`https://doc.deno.land/https/deno.land/x/oak@v6.3.1/mod.ts#ServerRequest`) to get its output in JSON. We then do a simple validation to check if the username and password are present in the request body, and at the bottom, we use the injected register method from the controller. We're wrapping it in a `try catch` so that we can return HTTP status code `400` if an error happens.

This should be enough for the web layer to be able to answer our request perfectly. Now, we just need to connect everything together.

Wiring the user controller with the web layer

We have created the fundamental pieces of the application. There's the business logic, there's the data accessing logic, and there's the web server to handle the request. The only thing that is missing is something that connects them. In this section, we'll instantiate the actual implementations of the interfaces we've defined and inject them into the content that's expecting them.

Go back into `src/index.ts`. Let's do something similar to what we did with the `museums` module. Here, we'll import the user repository and controller, instantiate them, and send the controller to the `createServer` function.

Follow the steps to do so:

1. In src/index.ts, import the user Controller and Repository from the user module and instantiate them, sending the necessary dependencies while doing so:

```
import {
  Controller as UserController,
  Repository as UserRepository,
    } from './users/index.ts';

...

const userRepository = new UserRepository();
const userController = new UserController({
  userRepository });
```

2. Send the user controller to the createServer function:

```
createServer({
  configuration: { port: 8080 },
  museum: museumController,
  user: userController
})
```

And with that, we're done! To finish this, let's run our application by running the following command:

```
$ deno run --allow-net src/index.ts
Application running at http://localhost:8080
```

Now, let's test the registered endpoint by making a request to /api/users/register with curl:

```
$ curl -X POST -d '{"username": "alexandrempsantos",
"password": "testpw" }' -H 'Content-Type: application/json'
http://localhost:8080/api/users/register
{"user":{"username":"alexandrempsantos","createdAt":"2020-10-
06T21:56:54.718Z"}}
```

As we can see, it's working and returning the contents of UserDto. Our main objective for this chapter has been achieved: we've created the user module and added an endpoint to register a user!

Summary

Our application went through a big change in this chapter!

We started by migrating our application from the standard library HTTP module to Oak. Not only did we migrate the logic to serve the app, but we also started to define some routes using Oak's router. We noticed that the application logic started to become simpler as Oak encapsulated part of the job that was done manually previously. We managed to migrate all the HTTP code from the standard library without having to change the business logic, which is a very good sign that we're doing well in terms of application architecture.

We kept moving and learned how to listen and handle events in an Oak application. As we started writing more code, we also became more familiar with Oak, understanding its functionalities, exploring its documentation, and experimenting with it.

Users are an important part of any application and with that in mind, we also spent a big part of this chapter focusing on them. We not only added users to our application but added it as a separate, self-contained module, alongside museums.

Once we'd developed the business logic for registering a user in the application, the need for a persistency layer for it was imminent. This meant we had to develop a user repository, which is responsible for creating users in the database. Here, we dived a little deeper and implemented a hash and salt mechanism to store the user's password on the database securely, while learning about a few Deno APIs in the process.

With the user business logic complete, we moved on to the part that was missing: the HTTP endpoint. We added the register route to our HTTP router and got everything set up with the help of Oak.

To wrap things up, we wired everything up again using dependency injection. Since all our modules' dependencies were based on interfaces, we easily injected the needed dependencies and got our code to work.

This chapter was a journey toward making our application more scalable and readable. We started by removing our DIY router code and moving it into Oak, and ended by us adding one big and important *business* entity – users. The latter also worked as a test for our architecture and to demonstrate how it can scale with different business domains.

In the next chapter, we'll keep iterating on the application by adding some interesting features. By doing this, we will complete the functionality we created here, such as user login, authorization, and persistence in a real database. Otherthings we'll tackle will include common API practices, such as basic logging and error handling.

Excited? So are we – let's go!

6
Adding Authentication and Connecting to the Database

In the previous chapter, we added an HTTP framework to our application, heavily simplifying our code. After that, we added the concept of users to the application and developed the register endpoint. In its current state, our application is already storing a couple of things, with the small gotcha that it's storing it in memory. We'll tackle this specific issue in this chapter.

Another concept that we've used while implementing oak (the HTTP framework of choice) was middleware functions. We'll start this chapter by learning what middleware functions are, and why they are one of the *standards* in pretty much all Node.js and Deno frameworks when it comes to reusing code.

We'll then use middleware functions and implement login and authorization. Adding to that, we will learn how to use middleware to add standard features such as request logging and timing to the application.

With our application very close to completeness in terms of requirements, we'll spend the rest of this chapter learning how to connect to a real persistence engine. For this book, we'll be using MongoDB. We'll use the abstractions we had previously built to make sure the transition is smooth. We'll then create a new users repository so that it can connect to a database the same way we can with an in-memory solution.

By the end of this chapter, we'll have a complete application with support for register and user login. After logging in, users can also get a list of museums. This is all done with the business logic from the HTTP and persistence implementation.

After this chapter, we'll only come back to the application to add tests and deploy it, thus completing the full cycle of building an application.

In this chapter, we'll cover the following topics:

- Using middleware functions
- Adding authentication
- Adding authorization with JWT
- Connecting with MongoDB

Let's get started!

Technical requirements

The code required for this chapter is available at the following GitHub link: `https://github.com/PacktPublishing/Deno-Web-Development/tree/master/Chapter06`.

Using middleware functions

If you have used any HTTP framework, be it JavaScript or otherwise, you are probably familiar with the concept of middleware functions. If you are not, then no worries – that's what we'll explain in this section.

Let's start with a definition borrowed from the Express.js documentation (`http://expressjs.com/en/guide/writing-middleware.html`):

> *"Middleware functions are functions that have access to the request object (req), the response object (res), and the next middleware function in the application's request-response cycle. The next middleware function is commonly denoted by a variable named next."*

Middleware functions intercept requests and have the ability to act on them. They can be used in many different use cases, as follows:

- Changing the request and response objects
- Ending the request-response life cycle (answering requests or skipping other handlers)
- Calling the next middleware function

Middleware functions are commonly used in tasks such as checking authentication tokens and automatically responding according to the result, logging requests, adding a specific header to a request, enriching the request object with context, and error handling, among other things.

We'll implement some of these examples in the application.

How does middleware work?

Middleware is processed as a stack, and each function can control the flow of the response with the ability to run the code before and after the rest of the stack executes.

In oak, middleware functions are registered by the `use` function. At this point, you might remember what we previously did with oak's router. What the `Router` object from oak does is create handlers for the registered routes and export two middleware functions with that behavior to be registered on the main application. These are called `routes` and `allowedMethods` (`https://github.com/PacktPublishing/Deno-Web-Development/blob/43b7f7a40157212a3afbca5ba0ae20f862db38c4/ch5/sections/2-2-handling-routes-in-an-oak-application/museums-api/src/web/index.ts#L38`).

To better understand middleware functions, we'll implement a couple of them. We'll do this in the next section.

Adding request timing via middleware

Let's add basic logging to our requests using some middleware. Oak middleware functions (`https://github.com/oakserver/oak#application-middleware-and-context`) receive two parameters. The first one is the context object, which is the same one that all the routes get, while the second one is the `next` function. This function can be used to execute other middleware in the stack, allowing us to control the application flow.

The first thing we'll do is add a middleware that adds the X-Response-Time header to the response. Follow these steps:

1. Go to src/web/index.ts and register a middleware that executes the rest of the stack by calling next.

    This adds a header to the response with the difference in milliseconds from the time the request started until it was handled:

    ```
    const app = new Application();
    app.use(async (ctx, next) => {
      const start = Date.now();
      await next();
      const ms = Date.now() - start;
      ctx.response.headers.set("X-Response-Time", `${ms}ms`);
    });
    ...
    app.use(apiRouter.routes());
    app.use(apiRouter.allowedMethods());
    ```

    This middleware should be added before any other .use calls; this way, all the other middleware functions will run once this has been executed.

    The first lines are executed before the route handler (and other middleware functions) starts handling the request. Then, the call to next makes sure the route handlers execute; only then is the rest of the middleware code executed, thus calculating the difference from the initial value and the current date and adding it as a header.

2. Execute the following code to get the server running:

    ```
    $ deno run --allow-net src/index.ts
    Application running at http://localhost:8080
    ```

3. Perform a request and check whether the desired header is there:

    ```
    $ curl -i http://localhost:8080/api/museums
    HTTP/1.1 200 OK
    content-length: 472
    x-response-time: 50ms
    content-type: application/json; charset=utf-8
    ```

And there we are! We have the `x-response-time` header there. Note that we've used the `-i` flag so that we're able to see the response headers on `curl`.

With that, we've used middleware functions after completely understanding them for the first time. We used them to control the flow of the application, by using `next`, and to enrich the request with a header.

The next thing we'll do is compose on the middleware we just created and add logic to log what request is being made to the server.

Adding request logging via middleware

Now that we have the logic to calculate the request timing we've already built, we're in a great place to add request logging to our application.

The final goal is to have every request that is made to the application logged to the console with its path, HTTP method, and the time it took to answer; something like the following example:

```
GET http://localhost:8080/api/museums - 65ms
```

We could, of course, do this individually per request, but since this is something that we want to do cross-application, we'll add it as a piece of middleware to the `Application` object.

The middleware we wrote in the previous section requires the handlers (and middleware functions) to run for it to add the response time (it calls the next function before executing part of the logic). We'll need to register the current logging middleware before the one we previously did, which added the timing to the request. Let's get started:

1. Go to `src/web/index.ts` and add the code for logging the request method, the path, and the timing to the console:

```
app.use(async (ctx, next) => {
  await next();
  const rt = ctx.response.headers.get("X-Response-Time");
  console.log(`${ctx.request.method} ${ctx.request.url}
    - ${rt}`);
});

...

app.use(apiRouter.routes());
app.use(apiRouter.allowedMethods());
```

Note how we're relying on the X-Response-Time header, which is going to be set by the previous middleware to log the request to the console. We're also using next to make sure all the handlers (and middleware functions) run before we log to the console. We need this specifically because the header is set by another piece of middleware.

2. Execute the following code to get the server running:

```
$ deno run --allow-net src/index.ts
Application running at http://localhost:8080
```

3. Perform a request to an endpoint:

```
$ curl http://localhost:8080/api/museums
```

4. Check that the server process is logging the request to the console:

```
$ deno run --allow-net src/index.ts
Application running at http://localhost:8080
GET http://localhost:8080/api/museums - 46ms
```

And with that, we have our middleware functions working together!

Here, we've registered middleware functions directly on the main application object. However, it is also possible to do this on specific oak routers by calling the same use method.

To give you an example, we'll register a middleware that will execute only on the / api routes. We will do the exact same thing we did previously, but instead of the Application object, we'll call the use method on the API Router object, as shown in the following example:

```
const apiRouter = new Router({ prefix: "/api" })
apiRouter.use(async (_, next) => {
  console.log("Request was made to API Router");
  await next();
}))
...
app.use(apiRouter.routes());
app.use(apiRouter.allowedMethods());
```

Middleware functions that want the application flow to proceed normally *must call the* next *function*. If this doesn't happen, the rest of the middleware in the stack and route handlers will not be executed, and thus the request will not be answered.

There's yet another way of using middleware functions: by directly adding them before the request handlers.

Imagine that we want to create a middleware that adds the X-Test header to some routes. We can either write that middleware on the application object or we can use it directly on the routes itself, as shown in the following code:

```
import { Application, Router, RouterMiddleware } from
   "../deps.ts";
...
const addTestHeaderMiddleware: RouterMiddleware = async (ctx,
   next) => {
   ctx.response.headers.set("X-Test", "true");
   await next();
}
apiRouter.get("/museums", addTestHeaderMiddleware, async (ctx)
   => {
   ctx.response.body = {
      museums: await museum.getAll()
   }
});
```

To get the preceding code working, we'll need to export the RouterMiddleware type in src/deps.ts:

```
export type { RouterMiddleware } from
   "https://deno.land/x/oak@v6.3.1/mod.ts";
```

With this middleware, whenever we want the X-Test header to be added, we just need to include addTestHeaderMiddleware before the route handler. It will execute before the handler's code. This is not exclusive to one piece of middleware, as multiple middleware functions can be registered.

And that's it for middleware functions!

We've learned the basics that allow us to create and share functionality by using this very common feature of web frameworks. We'll keep using them as we move into the next section, where we'll handle authentication, validate tokens, and authorize users.

Let's go and implement our application's authentication!

Adding authentication

In the previous chapter, we added the capability of creating new users to our application. This, by itself, is a cool feature, but it's not worth much if we can't use it for authentication. That's what we'll do here.

We'll start by creating the logic that checks whether a username and password combination is correct, and then we'll implement an endpoint to do that.

After this, we'll transition into the authorization topic by returning a token from the login endpoint, and later using that token to check if a user is authenticated.

Let's go step by step, starting with the business logic and persistency layer.

Creating the login business logic

It's already a practice of ours to, when writing new functionality, start with the business logic. We believe this is intuitive, as you think "business" and user first, and only then proceed into the technical details. That's what we'll do here.

We'll start by adding the login logic, back in `UserController`:

1. Before we start, let's add the `login` method to the `UserController` interface in `src/users/types.ts`:

    ```
    export type RegisterPayload = { username: string;
      password: string };
    export type LoginPayload = { username: string; password:
      string };

    export interface UserController {
      register: (payload: RegisterPayload) =>
        Promise<UserDto>;
      login: (
        { username, password }: LoginPayload,
    ```

```
    ) => Promise<{ user: UserDto }>;
}
```

2. Declare the `login` method on the controller; it should receive a username and a password:

```
public async login(payload: LoginPayload) {
}
```

Let's stop and think about what the flow of the login should be:

*   The user sends their username and password.
*   The application gets the user from their database by username.
*   The application encodes the user-sent password with the salt from the database.
*   The application compares both salted passwords.
*   The application returns a user and a token.

We won't worry about the token for now. However, the rest of the flow should set the requirements for the current section, helping us think about the code for the `login` method.

Just by looking at these requirements, we can understand that we'll need to have a method on `UserRepository` to get a user by username. Let's take a look at this.

3. In `src/users/types.ts`, add a `getByUsername` method to `UserRepository`; it should get a user from the database by username:

```
export interface UserRepository {
    create: (user: CreateUser) => Promise<User>;
    exists: (username: string) => Promise<boolean>
    getByUsername: (username: string) => Promise<User>
}
```

4. Implement the `getByUsername` method in `src/users/repository.ts`:

```
export class Repository implements UserRepository {
    ...
    async getByUsername(username: string) {
        const user = this.storage.get(username);
        if (!user) {
            throw new Error("User not found");
```

```
        }
        return user;
    }
}
```

Now, we can go back to `UserController` and use the recently created method to get a user from the database.

5. Use the `getByUsername` method from the repository inside the `login` method of `UserController`:

```
public async login(payload: LoginPayload) {
    try {
        const user = await
        this.userRepository.getByUsername(payload.username);
    } catch (e) {
        throw new Error("Username and password combination is
            not correct")
    }
}
```

Now that we have the user from the database, we'll have to compare its hash with the hashed password the user sent. We created a function called `hashPassword` in the previous chapter when we implemented the register logic, so let's use that.

6. Create a `comparePassword` method inside `UserController`.

It should receive a password and a `user` object. Then, it should compare the password that was sent by the user once it's been salted and hashed with what is stored in the database:

```
import {
    LoginPayload,
    RegisterPayload,
    User,
    UserController,
    UserRepository,
} from "./types.ts";
import { hashWithSalt } from "./util.ts"

...

private async comparePassword(password: string, user:
```

```
User) {
    const hashedPassword = hashWithSalt (password,
        user.salt);

    if (hashedPassword === user.hash) {
        return Promise.resolve(true);
    }

    return Promise.reject(false);
}
```

7. Use the `comparePassword` method on the `login` method of `UserController`:

```
public async login(payload: LoginPayload) {
    try {
        const user = await
            this.userRepository.getByUsername(payload.username);

        await this.comparePassword(payload.password, user);

        return { user: userToUserDto(user) };
    } catch (e) {
        console.log(e);
        throw new Error('Username and password combination is
            not correct')
    }
}
```

And with that, we have the `login` method working!

It receives a username and a password, gets a user by username, compares the hashed passwords, and returns the user if everything goes according to plan.

We should now be ready to implement the login endpoint – one that will use the login method we just created.

Creating the login endpoint

Now that we've created the business logic and data fetching logic, we can start using it in our web layer. Let's create the POST /api/login route, which should let the user log in with their username and password. Follow these steps:

1. In src/web/index.ts, create the login route:

```
apiRouter.post("/login", async (ctx) => {
})
```

2. Get the body of the request by using the request.body function (https://doc.deno.land/https/raw.githubusercontent.com/oakserver/oak/main/request.ts#Request) and then send the username and password to the login method:

```
apiRouter.post("/login", async (ctx) => {
  const { username, password } = await
    ctx.request.body().value;
  try {
    const { user: loginUser } = await user.login({
      username, password });
  } catch (e) {
    ctx.response.body = { message: e.message };
    ctx.response.status = 400;
  }
})
```

Note how we're handling the error by using a try-catch and returning the proper error code (400 Bad Request) if things didn't go well.

3. If the login succeeds, it should return our user:

```
...
const { user: loginUser } = await user.login({ username,
  password });
ctx.response.body = { user: loginUser };
ctx.response.status = 201;
...
```

With that, we should have all it takes to log a user in! Let's try it out.

4. Run the application by running the following command:

```
$ deno run --allow-net src/index.ts
Application running at http://localhost:8080
```

5. Perform a request to register the user at /api/users/register, and then try to log in with the created user at /api/login:

```
$ curl -X POST -d '{"username": "asantos00", "password":
"testpw" }' -H 'Content-Type: application/json' http://
localhost:8080/api/users/register
{"user":{"username":"asantos00","createdAt":"2020-10-
19T21:30:51.012Z"}}
```

6. Now, try to log in with the created user:

```
$ curl -X POST -d '{"username": "asantos00", "password":
"testpw" }' -H 'Content-Type: application/json' http://
localhost:8080/api/login
{"user":{"username":"asantos00","createdAt":"2020-10-
19T21:30:51.012Z"}}
```

And it works! We're creating the user on the registry and are able to log in with them afterward.

In this section, we learned how to add authentication logic to our application and implemented the login method, which allows users to log in with a registered user.

In the next section, we'll learn how to use this authentication we've created to get a token that will allow us to handle authorization. We'll make the museums route only available for authenticated users, instead of being publicly available. For this, we will need to develop the authorization feature. Let's jump in!

Adding authorization with JWT

We now have an application that allows us to log in and return the logged in user. However, if we want to use the login in any API, we'll have to create an authorization mechanism. This mechanism should enable the users of the API to authenticate, get a token, and use that token to identify themselves and access resources.

We're doing this as we want to close part of the application's routes so that they're only available to authenticated users.

We'll develop what's needed to integrate with token authentication by using **JSON Web Tokens (JWT)**, which is pretty much a standard in APIs nowadays.

If you are not familiar with JWT, I'll leave you with an explanation from `jwt.io`:

> *"JSON Web Tokens are an open, industry standard RFC 7519 method for representing claims securely between two parties."*

It is mainly used when you want your clients to connect to an authentication service, and them provide your servers with the ability to verify if that authentication was issued by a service that you trust.

To avoid the risk of repeating what has already been very well-explained by `jwt.io`, I'll leave you with a link that explains what this standard is perfectly: `https://jwt.io/introduction/`. Make sure to give it a read; I'm sure you have all it takes to understand about how we'll be using it next.

In this section, and due to the scope of this book, we will not implement the whole logic to generate and validate JWT tokens. The code for that is available in this book's GitHub repository (`https://github.com/PacktPublishing/Deno-Web-Development/tree/master/Chapter06/jwt-auth`).

What we will do here is integrate our current application with a module that has functions for generating and validating JWT tokens, which is what matters for our application. Then, we'll use that token to decide whether we're letting the user access the museums route.

Let's go!

Returning a token from login

In the previous section, we implemented the login functionality. We developed some logic that validates the combination of username and password, returning the user if it succeeds.

In order to authorize a user and let them access private resources, we need to know who the authenticated user is. A common way to do this is via a token. There are various ways we can do this. They are alternatives such as basic HTTP authentication, sessions tokens, JWT tokens, and so on. We chose JWT as we believe it is a heavily used solution that you might have encountered in the industry. Don't worry if you haven't; it is simple enough to grasp.

The first thing we will need to do is return a token to the user when they log in. Our `UserController` will have to return that token in conjunction with `userDto`.

In the provided `jwt-auth` module (`https://github.com/PacktPublishing/Deno-Web-Development/tree/master/Chapter06/jwt-auth`), you can check that we're exporting a Repository.

If we access the documentation, using Deno's documentation website at `https://doc.deno.land/https/raw.githubusercontent.com/PacktPublishing/Deno-Web-Development/master/Chapter06/jwt-auth/repository.ts`, we can see that it exports two methods: `getToken` and `generateToken`.

Reading the method's documentation, we can understand that one gets a token for a user ID, and that the other generates a new token, respectively.

Let's use this method to generate a new token in our login use case by following these steps:

1. Start by adding the token to the return type of `UserController` in `src/users/types.ts`:

    ```
    export interface UserController {
      register: (payload: RegisterPayload) =>
        Promise<UserDto>
      login: ({ username, password }: LoginPayload) =>
        Promise<{ user: UserDto, token: string }>
    }
    ```

    Now, we need to make sure `UserController` knows how to return a token. Looking at its logic, we can see that it should be able to delegate that responsibility by calling a method that will return that token.

    From the previous chapters, we know that we don't want to import our dependencies directly; we'd rather have them injected into our `constructor`. That's what we'll do here.

    Another thing we know is that we want to use this "third-party module" that deals with authentication. We'll need to add it to our dependencies file.

2. Go to `src/deps.ts` and add the export for the `jwt-auth` module, running `deno cache` to update the lock file and download the dependencies:

    ```
    export type {
      Algorithm,
    } from "https://raw.githubusercontent.com/PacktPublishing/
    Deno-Web-Development/master/Chapter06/jwt-auth/mod.ts";
    ```

```
export {
  Repository as AuthRepository,
} from "https://raw.githubusercontent.com/
PacktPublishing/
  Deno-Web-Development/master/Chapter06/jwt-auth/mod.ts";
```

3. Use the `AuthRepository` type to define the `UserController` constructor's dependencies:

```
import { AuthRepository } from "../deps.ts";

interface ControllerDependencies {
  userRepository: UserRepository
  authRepository: AuthRepository
}

export class Controller implements UserController {
  userRepository: UserRepository;
  authRepository: AuthRepository;

  constructor({ userRepository, authRepository }:
    ControllerDependencies) {
    this.userRepository = userRepository;
    this.authRepository = authRepository;
  }
```

Now, it's time to start using `authRepository`, which we've just imported. We previously discovered that it exposes a `generateToken` method, which will be of use to the login of `UserController`.

4. Go to the login method in `src/users/controller.ts` and use the `generateToken` method from `authRepository` to get a token and return it:

```
public async login(payload: LoginPayload) {
    try {
      const user = await
        this.userRepository.getByUsername
          (payload.username);
```

```
        await this.comparePassword(payload.password, user);

        const token = await
          this.authRepository.generateToken(user.username);

        return { user: userToDto(user), token };
      ...
```

Here, we're using `authRepository` to get a token.

If we try to run this code, we know it will fail. In fact, we just need to open `src/index.ts` to see our editor's warnings. It is complaining that we're not sending `authRepository` to `UserController`, and we should.

5. Go back to `src/index.ts` and instantiate `AuthRepository` from `jwt-auth`:

```
    import { AuthRepository } from "./deps.ts";
    ...
    const authRepository = new AuthRepository({
      configuration: {
        algorithm: "HS512",
        key: "my-insecure-key",
        tokenExpirationInSeconds: 120
      }
    });
```

You can also check by the module's documentation, since it requires a `configuration` object to be sent with three properties; that is, `algorithm`, `key`, and `tokenExpirationInSeconds`.

`key` should be a secret value that is used to create and validate the JWT, `algorithm` is the crypto algorithm that the token will be encoded with (HS256, HS512, and RS256 supported), and `tokenExpirationInSeconds` is the time it takes for the token to expire.

Regarding the values that are secret and shouldn't live in the code, such as the `key` variable that we just mentioned, we'll learn how to handle them in the next chapter, where we'll talk about application configuration.

We now have an instance of `AuthRepository`! We should be able to send it to our `UserController` and get it working.

6. In `src/index.ts`, send `authController` into the `UserController` constructor:

```
const userController = new UserController({
  userRepository, authRepository });
```

Now, you should be able to run the application!

Now, if you create a few requests to test it, you'll notice that the POST /login endpoint is still not returning the token. Let's fix this!

7. Go to `src/web/index.ts` and, on the `login` route, make sure we're getting `token` returned from the `login` method present in the response:

```
apiRouter.post("/login", async (ctx) => {
  const { username, password } = await
    ctx.request.body().value;
  try {
    const { user: loginUser, token } = await user.login({
      username, password });
    ctx.response.body = { user: loginUser, token };
    ctx.response.status = 201;
  } catch (e) {
    ctx.response.body = { message: e.message };
    ctx.response.status = 400;
  }
})
```

We're almost done! We managed to finish our first objective: having the `login` endpoint return a token.

The next thing we want to implement is the logic that makes sure that a user is sending a token whenever they're trying to access an authenticated route.

Let's go and finish the authorization logic.

Making an authenticated route

Having the capacity to get users a token, we now want a guarantee that only logged in users are able to access the museums route.

Users will have to send the token in the `Authorization` header, as the JWT token standard defines. If the token is invalid or not present, the user should be presented with a `401 Unauthorized` status code.

Validating the token that's been sent by users on the request is a nice use case for middleware functions.

In order to do this, and since we're using `oak`, we'll be using a third-party module named `oak-middleware-jwt`. This is nothing more than a middleware that automatically validates the JWT token, based on a key, and provides functionality that will be useful to us.

You can check its documentation at `https://nest.land/package/oak-middleware-jwt`.

Let's use this middleware in our web code to make the museums route only available to authenticated users. Follow these steps:

1. Add `oak-middleware-jwt` to the `deps.ts` file and export the `jwtMiddleware` function:

    ```
    export {
      jwtMiddleware,
    } from "https://x.nest.land/
      oak-middleware-jwt@2.0.0/mod.ts";
    ```

2. Back in `src/web/index.ts`, use `jwtMiddleware` in the museums route, sending the key and the algorithm there.

    Do not forget what we mentioned in previous section – that middleware functions can be used in any route by sending it before the route handler:

    ```
    import { Application, jwtMiddleware, Router } from
      "../deps.ts";
    ...
    const authenticated = jwtMiddleware
      ({ algorithm: "HS512", key: "my-insecure-key" })
    apiRouter.get("/museums", authenticated, async (ctx) => {
      ctx.response.body = {
        museums: await museum.getAll()
      }
    });
    ```

You might have noticed that we're sending the algorithm and key here again. Even though it currently works, this is a point where our application can fail. Imagine that we change the configurations we currently have in `src/index.ts` and forget to change this.

This is exactly why we should extract this and expect it as a parameter to the `createServer` function.

3. Add `authorization` as a parameter inside `configuration` in the `createServer` function:

```
import { Algorithm, Application, jwtMiddleware, Router }
  from "../deps.ts";


interface CreateServerDependencies {
  configuration: {
    port: number,
    authorization: {
      key: string,
      algorithm: Algorithm
    }
  },
  museum: MuseumController,
  user: UserController
}


export async function createServer({
  configuration: {
    port,
    authorization
  },
...
```

Note how we're importing the `Algorithm` type from the `deps.ts` file, which exports it from the `jwt-auth` module. We're doing this so that we can ensure, via types, that the algorithms that are sent are only the supported ones.

4. Now, still in `src/web/index.ts`, use the `authorization` params to send the values that will be injected to `jwtMiddleware`:

```
const authenticated = jwtMiddleware(authorization)
```

The only thing we're missing is the ability to actually send the `authorization` value to the `createServer` function.

5. In `src/index.ts`, extract the auth configuration into a variable so that we can reuse it:

```
import { AuthRepository, Algorithm } from "./deps.ts";
...
const authConfiguration = {
  algorithm: "HS512" as Algorithm,
  key: "my-insecure-key",
  tokenExpirationInSeconds: 120
}
const authRepository = new AuthRepository({
  configuration: authConfiguration
});
```

6. Let's reuse that same variable to send the required parameters to `createServer`:

```
createServer({
  configuration: {
    port: 8080,
    authorization: {
      key: authConfiguration.key,
      algorithm: authConfiguration.algorithm
    }
  },
  museum: museumController,
  user: userController
})
```

And we're done! Let's test our application and see if it works as expected.

Note that the desired behavior is that only an authenticated user can access the museums route and see all the museums.

7. Let's run the application by running the following command:

```
$ deno run --allow-net src/index.ts
Application running at http://localhost:8080
```

8. Let's register a user so that we can log in:

```
$ curl -X POST -d '{"username": "asantos00", "password":
"testpw1" }' -H 'Content-Type: application/json'
http://localhost:8080/api/users/register
{"user":{"username":"asantos00","createdAt"
:"2020-10-27T19:14:01.984Z"}}
```

9. Now, let's log in so that we can get our token:

```
$ curl -X POST -d '{"username": "asantos00", "password":
"testpw1" }' -H 'Content-Type: application/json'
http://localhost:8080/api/login
{"user":{"username":"asantos00","create
dAt":"2020-10-27T19:14:01.984Z"},"token"
:"eyJhbGciOiJIUzUxMiIsInR5cCI6IkpXVCJ9.
eyJpc3MiOiJtdXNldW1zIiwiZXhwIjoxNjAzODI2NTEzLCJ1c2VyIjoi
YXNhbnRvczAwIn0.XV1vaHDpTu2SnavFla5q8eIPKCRIfDw_Kk-j8gi1
mqcz5UN3sVnk61JWCapwlh0IJ46fJdc7cw2WoMMIh-ypcg"}
```

10. Finally, let's try to access the museums route with the token that was returned from the previous request:

```
$ curl -H 'content-type: application/json' -H
'Authorization: Bearer eyJhbGciOiJIUzUxMiIsInR5cCI6IkpXV
CJ9.eyJpc3MiOiJtdXNldW1zIiwiZXhwIjoxNjAzODI2NTEzLCJ1c2Vy
IjoiYXNhbnRvczAwIn0.XV1vaHDpTu2SnavFla5q8eIPKCRIfDw_Kk-
j8gi1mqcz5UN3sVnk61JWCapwlh0IJ46fJdc7cw2WoMMIh-ypcg'
http://localhost:8080/api/museums
{"museums":[{"id":"fixture-1","name":"Most beautiful
museum in the world","description":"One I really
like","location":{"lat":"12345","lng":"54321"}}]}
```

We've got it working!

Note how we're sending the Authentication header with Bearer as a prefix, as specified by the JWT specification.

11. Just to make sure it works as expected, let's try to do the same request without the Authorization header, expecting an unauthorized response:

```
$ curl -i -H 'content-type: application/json'
http://localhost:8080/api/museums
HTTP/1.1 401 Unauthorized
content-length: 21
content-type: text/plain; charset=utf-8
```

And we get the expected response! Note how we're using the -i flag with curl so that it logs the request status code and headers.

That's it! With that, we've managed to make a route only accessible by authenticated users. This is something very common in any application that contains users.

If we were to go deeper into this, we could have explored the JWT refreshToken, or even how to read the user information from the JWT token, but that's outside the scope of this book. This is something I'll let you explore on your own.

In this section, we achieved our goal and looked at many different parts of an API.

There's one thing missing, though: a connection with a real persistence engine. That's what we're going to do next – connect our application to a NoSQL database!

Connecting to MongoDB

So far, we've implemented an application that lists museums, and contains users, allowing them to authenticate. These features are already in place, but they all have a catch: they're all running against an in-memory database.

We've decided to do it this way for the sake of simplicity. However, since most of our implementation doesn't depend on the delivery mechanism, it shouldn't change much if the database changes.

As you might have guessed by this section's title, we'll learn how to move one of the application entities to the database. We'll leverage the abstractions we've created in order to do this. The process will be very similar to all the entities, and thus we've decided on learning how to connect to a database just for the users' module.

Later, if you are curious about how this would work if all the applications were connected to the database, you'll have the opportunity to check this book's GitHub repository.

To make sure we're all running against a similar database, we'll use MongoDB Atlas. Atlas is a product that provides a free MongoDB cluster that we can use to connect our application.

If you are not familiar with MongoDB, there's here's a "one-sentence explanation" from their website (`https://www.mongodb.com/`). Feel free to go there and learn a little more about it:

> *"MongoDB is a general purpose, document-based, distributed database built for modern application developers and for the cloud era."*

Ready? Let's go!

Creating a User MongoDB repository

Our current `UserRepository` is the module that's responsible for connecting the user to the database. This is the one we want to change in order to make our application connect with a MongoDB instance, instead of an in-memory database.

We'll go through the steps of creating the new MongoDB repository, exposing it to the world, and connecting the rest of our application to it.

Let's start by creating the space for a new users repository to exist, by reorganizing the users module's internal folder structure.

Rearranging our users module

Our users module was initially thought to have a single repository, and thus it doesn't have a folder for it; it's just a single `repository.ts` file. Now that we're thinking of adding more ways our users can be saved to a database, we need to create it.

Remember when we first talked about architecture, and mentioned that it would keep evolving? That's what's happening here.

Let's rearrange the users module so that it can handle multiple repositories and add a MongoDB repository, following the `UserRepository` interface we previously created:

1. Create a folder named `repository` inside `src/users` and move the actual `src/users/repository.ts` there, renaming it `inMemory.ts`:

```
└── src
    ├── museums
    ├── users
    │   ├── adapter.ts
```

```
|       ├── controller.ts
|       ├── index.ts
|       ├── repository
|       |    ├── inMemory.ts
|       ├── types.ts
|       └── util.ts
```

2. Remember to fix the module imports inside `src/users/repository/inMemory.ts`:

```
import { User, UserRepository } from "../types.ts";
import { generateSalt, hashWithSalt } from "../util.ts";
```

3. To keep the application running, let's go to `src/users/index.ts` and export the correct repository:

```
export { Repository } from './repository/inMemory.ts'
```

4. Now, let's create our MongoDB repository. Call it `mongoDb.ts` and put it inside the `src/users/respository` folder:

```
import { UserRepository } from "../types.ts";

export class Repository implements UserRepository {
    storage

    async create(username: string, password: string) {
    }

    async exists(username: string) {
    }

    async getByUsername(username: string) {
    }
}
```

Make sure it implements the `UserRepository` interface we defined previously.

This is where all the fun starts! Now that we have the MongoDB repository, we'll start writing it and connecting our application to it.

Installing the MongoDB client library

We already have a list of methods that our repository needs to have implemented. By following the interface, we can guarantee that our application will work, regardless of the implementation.

There's one thing that we know for sure, as we don't want to keep reinventing the wheel: we're going to use a third-party package to handle the connection with MongoDB.

We will use the deno-mongo package for this (https://github.com/manyuanrong/deno_mongo).

> **Important note**
> Deno's MongoDB driver uses the Deno plugins API, which is still in unstable mode. This means that we will have to run our application with the --unstable flag. As it is currently using APIs that are not yet considered stable, this shouldn't be used in production yet.

Let's have a look at the documentation's example, where a connection to a MongoDB database is established:

```
import { MongoClient } from
  "https://deno.land/x/mongo@v0.13.0/mod.ts";

const client = new MongoClient();
client.connectWithUri("mongodb://localhost:27017");

const db = client.database("test");
const users = db.collection<UserSchema>("users");
```

Here, we can see that we will need to create a MongoDB client and connect it to a database, using a connection string that contains the host (which might contain the host's username and password).

After that, we need to get the client to access a specific database (test, in this example). Only then can we have the handler that will let us interact with the collection (users, in this example).

First things first, let's add `deno-mongo` to our dependencies list:

1. Go to your `src/deps.ts` file and add the exports for `MongoClient` there:

    ```
    export { MongoClient } from
        "https://deno.land/x/mongo@v0.13.0/mod.ts";
    ```

2. Now, make sure you run the `cache` command to install the modules. We'll have to run it with the `--unstable` flag as the plugin we're installing requires unstable APIs on its installation too:

    ```
    $ deno cache --lock=lock.json --lock-write --unstable
    src/deps.ts
    ```

With that, we have updated the `deps.ts` file with the package we just installed!

Let's move on and actually use this package to develop our repository.

Developing the MongoDB repository

In the example we got from the documentation, we learned how to connect to the database and create the handlers for the user's collection that we want. We know that our repository needs to have access to this handler so that it can interact with the collection.

Again, we could create the MongoDB client directly inside the repository, but that would make it impossible for us to test that repository without trying to connect to MongoDB.

As we want the dependencies to be injected into the modules as much as possible, we'll pass the MongoDB client into our repository via its constructor, which is something very similar to what we did in other parts of the code.

Let's go back to our MongoDB repository and do this by following these steps:

1. Create the `constructor` method inside the MongoDB repository.

    Make sure it receives an object with a property named `storage` of the `Database` type, which is exported by the `deno-mongo` package:

    ```
    import { User, UserRepository } from "../types.ts";
    import { Database, Collection } from '../../deps.ts';

    interface RepositoryDependencies {
      storage: Database,
    }
    ```

```
export class Repository implements UserRepository {
  storage: Collection<User>

  constructor({ storage }: RepositoryDependencies) {
    this.storage = storage.collection<User>('users');
  }
  ...
```

Note that we're expecting a database, and then calling the `collection` method on it, to get access to the users' collection. Once we've done that, we must set it to our `storage` class property. Both the method and the type require a generic to be passed in. This should represent the type of object present in that collection. In our case, it is the `User` type.

2. Now, we have to go into the `src/deps.ts` file and export the `Database` and `Collection` types from `deno-mongo`:

```
export { MongoClient, Collection, Database } from
  "https://deno.land/x/mongo@v0.13.0/mod.ts";
```

Now, it's just a matter of developing the methods to satisfy the `UserRepository` interface.

These methods will be very similar to the ones we developed for the in-memory database, with the difference that we're now interacting with a MongoDB collection instead of the JavaScript Map we were using previously.

Now, we just need to implement some methods that will create, verify the existence of a user, and get it by username. These methods are available in the plugin documentation and very closely mimic MongoDB's native API.

This is what the final class is going to look like:

```
import { CreateUser, User, UserRepository } from
  "../types.ts";
import { Collection, Database } from "../../deps.ts";

export class Repository implements UserRepository {
  storage: Collection<User>
  constructor({ storage }: RepositoryDependencies) {
    this.storage = storage.collection<User>("users");
  }
```

```
  async create(user: CreateUser) {
    const userWithCreatedAt = { ...user, createdAt: new Date()
}
    this.storage.insertOne({ ...user })
    return userWithCreatedAt;
  }
  async exists(username: string) {
    return Boolean(await this.storage.count({ username }));
  }
  async getByUsername(username: string) {
    const user = await this.storage.findOne({ username });
    if (!user) {
      throw new Error("User not found");
    }
    return user;
  }
}
```

We've highlighted the methods that use the deno-mongo plugin to access the database. Note how the logic is very similar to what we've previously done. We're adding the created at date to the create method, and then calling the create method from mongo. In the exists method, we're calling the count method, and converting it into a boolean. For the getByUsername method, we're using the findOne method from the mongo library, sending the username in.

If you have any questions about how these APIs can be used, please check out deno-mongo's documentation (https://github.com/manyuanrong/deno_mongo).

Connecting the application to MongoDB

Now, in order to expose the MongoDB repository that we've created, we need to go into src/users/index.ts and expose it as Repository (delete the highlighted line):

```
export { Repository } from "./repository/mongoDb.ts";
export { Repository } from "./repository/inMemory.ts";
```

We should now have our editor and typescript compiler complaining that we're not sending the correct dependencies into UserRepository at the moment of its instantiation on src/index.ts, which is true. So, let's go there and fix it.

Before we send the database client into `UserRepository`, it needs to be instantiated. By looking at the documentation for `deno-mongo`, we can read the following example:

```
const client = new MongoClient();
client.connectWithUri("mongodb://localhost:27017");
```

We aren't connecting with the localhost, so we'll need to change the connection URI later.

Let's follow the documentation's example and write the code for connecting to a MongoDB instance. Follow these steps:

1. After adding the export of `MongoClient` to the `src/deps.ts` file, import it in `src/index.ts`:

    ```
    import { MongoClient } from "./deps.ts";
    ```

2. Then, call `connectWithUri`:

    ```
    const client = new MongoClient();
    client.connectWithUri("mongodb://localhost:27017");
    ```

3. After that, get a database handler by calling the `database` method on the client:

    ```
    const db = client.database("getting-started-with-deno");
    ```

And that should be all we need in order to connect to MongoDB. The only thing missing is the code for sending the database handler into the `UserRepository`. So, let's add this:

```
const client = new MongoClient();
client.connectWithUri("mongodb://localhost:27017");
const db = client.database("getting-started-with-deno");
...
const userRepository = new UserRepository({ storage: db });
```

No warnings should be visible and we should be able to run our application now!

However, we still do not have a database to connect to. We'll look at this next.

Connecting to a MongoDB cluster

Now, we need to connect to a real MongoDB instance. Here, we'll be using a service called Atlas. Atlas is a service from MongoDB that provides a cloud MongoDB database. Their free tier is quite generous and works well for our application. Create an account there. Once you've done that, we can create a MongoDB cluster.

> **Important Note**
>
> If you have any other MongoDB instance, local or remote, feel free to use it by skipping the next paragraph and going directly to inserting the database URI into the code.

The following link contains all the instructions needed to create a cluster: `https://docs.atlas.mongodb.com/tutorial/create-new-cluster/`.

Once the cluster has been created, we also need to create a user that has access to it. Go to `https://docs.atlas.mongodb.com/tutorial/connect-to-your-cluster/index.html#connect-to-your-atlas-cluster` to learn how to get the connection string.

It should look something like the following:

```
mongodb+srv://<username>:<password>@clustername.mongodb.net/
   test?retryWrites=true&w=majority&useNewUrlParser=
      true&useUnifiedTopology=true
```

Now that we have the connection string, we just need to pass it to the code we created previously in `src/index.ts`:

```
const client = new MongoClient();
client.connectWithUri("mongodb+srv://<username>:<password>
   @clustername.mongodb.net/test?retryWrites=true&w=
      majority&useNewUrlParser=true&useUnifiedTopology=true");
const db = client.database("getting-started-with-deno");
```

And that should be all we need to get our application running. Let's do this!

Keep in mind that since we're using the plugins API to connect to MongoDB, and it's still unstable, the following permissions are needed together with the `--unstable` flag:

```
$ deno run --allow-net --allow-write --allow-read --allow-
plugin --allow-env --unstable src/index.ts
Application running at http://localhost:8080
```

Now, to test that our `UserRepository` is working and connected to the database, let's try to register and log in and see if it works:

1. Perform a POST request to `/api/users/register` to register our user:

   ```
   $ curl -X POST -d '{"username": "asantos00", "password":
   "testpw1" }' -H 'Content-Type: application/json' http://
   localhost:8080/api/users/register
   {"user":{"username":"asantos00","createdAt":"2020-11-
   01T23:21:58.442Z"}}
   ```

2. Now, to make sure we are connecting to permanent storage, we can stop the application and run it again, before trying to log in:

   ```
   $ deno run --allow-net --allow-write --allow-read
   --allow-plugin --allow-env --unstable src/index.ts
   Application running at http://localhost:8080
   ```

3. Now, let's log in with that same user we just created:

   ```
   $ curl -X POST -d '{"username": "asantos00", "password":
   "testpw1" }' -H 'Content-Type: application/json' http://
   localhost:8080/api/login
   {"user":{"username":"asantos006"},"token":
   "eyJhbGciOiJIUzUxMiIsInR5cCI6IkpXVCJ9.eyJpc3MiOiJtdXN1
   dW1zIiwiZXhwIjoxNjA0MjczMDQ1LCJ1c2VyIjoiYXNhbnRvczAwNi
   J9.elY48It-DHse5sSszCAWuE2PzNkKiPsMIvif4v5klY1URq0togK
   84wsbSskGAfe5UQsJScr4_0yxqnrxEG8viw"}
   ```

And we have our response! We managed to connect the application that was previously connected to an in-memory database to a real MongoDB database. If you used MongoDB, you can view the users that were created there on the Atlas interface by going to the **Collections** menu.

Did you notice how we didn't need to touch any of our business or web logic just to change the persistency mechanism? This proves that the layers and abstractions we initially created are now paying off, by allowing decoupling between different parts of the application.

With that, we have completed this chapter and migrated our users to a real database. We could do the same for the other modules, but that would be mostly the same thing and will not add much to your learning experience. I'd like to challenge you on writing the other modules' logic so that it can connect to MongoDB.

If you want to skip this but you're curious about what it will look like, then take a look at this book's GitHub repository.

Summary

This chapter pretty much wraps up our application in terms of logic. We'll come back later in *Chapter 8, Testing – Unit and Integration*, to add tests and the single feature that we're missing – the ability to rate museums. However, most of this has already been done. In its current state, we have an application that has its domains divided into modules that can be used by themselves and don't depend on each other. We believe we achieved something that is both easy to navigate in the code and extendable.

This concludes the process of constantly reworking and refining the architecture, managing dependencies, and tweaking logic to make sure code is as decoupled as possible, and as easy to change in the future as possible. While doing all of this, we managed to create an application with a couple of features, trying to go around industry standards at the same time.

We started this chapter by learning about middleware functions, something we'd previously used, even though we still hadn't learned about them. We understood how they work, and how they can be leveraged to add logic across applications and routes. To get a little more concrete, we went into specific examples and finished by implementing a few of them in the application. Here, we added common capabilities such as basic logging and request timing.

Then, we went on to finish our journey on authentication. After adding users and registration in the previous chapter, we started by implementing the capability to authenticate. We relied on an external package to manage our JWT tokens, which we used later for our authorization mechanism. After providing our users with a token, we had to make sure that the token was valid and only then let the user access the application. We added an authenticated route to the museums route, making sure it can only be accessed by authenticated users. Once again, middleware was used to check the token's validity and answer the request on error cases.

We wrapped this chapter up by adding one more feature to the application: a connection to a real database. Before we did this, all our application modules were relying on an in-memory database. Here, we moved one of the modules, users, to a MongoDB instance. To do this, we leveraged the layers we previously created to separate business logic from our persistence and delivery mechanism. Here, we created and implemented what we called the MongoDB repository, ensuring the application is running smoothly but with a real persistence mechanism. We used MongoDB Atlas for this example.

In the next chapter, we'll add a couple more things to our web application, namely the capability to manage secrets and configurations outside code, a well-known best practice. We'll also explore the possibilities of Deno when it comes to running code in the browser, among other things. The next chapter will wrap up this part of this book; that is, building the features of the application. Let's go!

7
HTTPS, Extracting Configuration, and Deno in the Browser

In the previous chapter, we pretty much wrapped up our application's features. We added authorization and persistence, ending up with an application connected to a MongoDB instance. In this chapter, we'll focus on some known best practices that are standard in production applications: basic security practices and dealing with configuration.

First, we'll add a couple of basic security features to our **application programming interface** (**API**), starting with **Cross-Origin Resource Sharing** (**CORS**) protection, to enable the filtering of requests based on their origin. Then, we'll learn how to enable **HyperText Transfer Protocol Secure** (**HTTPS**) in our application so that it supports encrypted connections. This will allow users to perform requests to the API using a secure connection.

Until now, we've used a few secret values, but we weren't concerned about having them in the code. In this chapter, we'll extract the configuration and secrets so that they don't have to live in the code base. We'll then learn how we can have them safely stored and injected in the application. This way, we can be sure that those values are kept a secret and are not present in the code. By doing this, we'll also enable different deployments with different configurations.

Moving forward, we'll explore the capabilities enabled by one specific Deno feature: the ability to compile and run code in the browser. By using Deno's compatibility with ECMAScript 6 (supported by modern browsers), we'll share code between the API and the frontend, enabling a whole new world of possibilities.

Leveraging this specific feature, we'll explore one specific scenario: building a JavaScript client for the API. This client will be built using the same types and parts of code that also run on the server, and we'll explore the benefits provided by that.

This chapter wraps up the *Building an application* section of this book, whereby we built an application step by step, adding some common application features with an incremental approach. While learning, we also made sure this application was as close to real as possible for an introductory book. This enabled us to learn about Deno, many of its APIs, and some community packages while we created a functional application.

By the end of this chapter, you'll be familiar with the following topics:

- Enabling CORS and HTTPS
- Extracting configuration and secrets
- Running Deno code in the browser

Technical requirements

All the code files needed for this chapter can be found at the following GitHub link:

```
https://github.com/PacktPublishing/Deno-Web-Development/tree/
master/Chapter07/sections
```

Enabling CORS and HTTPS

CORS protection and HTTPS support are two things considered critical in any running production application. This section will explain how can we add them to the application that we're building.

There are many other security practices that can be added to any API. As those aren't Deno specifics and deserve a book by themselves, we decided to focus on these two elements.

We'll begin by learning about CORS and how can we leverage `oak` and the middleware function feature we know in order to do it. Then, we'll learn how can we also use a self-signed certificate and make our API handle secure HTTP connections.

Let's go, starting with CORS.

Enabling CORS

If you are not familiar with CORS, it is a mechanism that enables a server to indicate to browsers which origins they should allow resource loading from. When the application is running on the same domain as the API, CORS is not even necessary, as the name directly makes explicit.

I'll provide you with the following quote from **Mozilla Developer Network** (**MDN**), explaining CORS:

> *"Cross-Origin Resource Sharing (CORS) is an HTTP-header based mechanism that allows a server to indicate any other origins (domain, protocol, or port) than its own from which a browser should permit loading of resources. CORS also relies on a mechanism by which browsers make a "preflight" request to the server hosting the cross-origin resource, in order to check that the server will permit the actual request. In that preflight, the browser sends headers that indicate the HTTP method and headers that will be used in the actual request."*

To give you a more concrete example, imagine you have an API running at `the-best-deno-api.com` and you want to handle requests made from `the-best-deno-client.com`. Here, you'll want your server to have CORS enabled for the `the-best-deno-client.com` domain.

If you don't have it enabled, the browser will make a preflight request to your API (using the `OPTIONS` method), and the response to this request will not have an `Access-Control-Allow-Origin: the-best-deno-client.com` header, causing the request to fail and the browser to prevent further requests.

We'll learn how we can enable this mechanism in our application, allowing requests to be made from `http://localhost:3000` in the example that follows.

As our application is using the `oak` framework, we'll learn how to do it with this framework. However, this is very similar to any other HTTP framework. We basically want to add a middleware function that handles requests and verifies their origins against a list of allowed domains.

We'll use a community package called `cors` (`https://deno.land/x/cors@` `v1.2.1`), but the implementation is quite simple. If you're curious about what it does, take a look at `https://deno.land/x/cors@v1.2.1/oakCors.ts`, as the code is quite straightforward.

> **Important note**
>
> We'll use the code we created in the previous chapter to start this implementation. This is available at `https://github.com/` `PacktPublishing/Deno-Web-Development/tree/master/` `Chapter06/sections/4-connecting-to-mongodb/` `museums-api`. You can also have a look at the finished code for this section here:
>
> `https://github.com/PacktPublishing/Deno-Web-` `Development/tree/master/Chapter07/sections/3-deno-` `on-the-browser/museums-api`

Here, we'll add the `cors` package to our application, together with our own list of allowed domains. The end goal is that we can perform requests from a trusted website to this API.

Let's do it. Proceed as follows:

1. Install the `cors` module by updating the `deps` file (check *Chapter 3, The Runtime and Standard Library,* for reference on how to do this). The code can be seen in the following snippet:

```
export { oakCors } from
    "https://deno.land/x/cors@v1.2.1/oakCors.ts";
```

2. Next, run the `cache` command to update the `lock` file, as follows:

```
$ deno cache --lock=lock.json --lock-write --unstable
src/deps.ts
```

3. Import `oakCors` on `src/web/index.ts` and register it on the application, before the router is registered, as follows:

```
import { Algorithm, oakCors } from "../deps.ts"
…
app.use(
  oakCors({ origin: ['http://localhost:3000'] })
);
const apiRouter = new Router({ prefix: "/api" });
```

Note how we're using the `oakCors` middleware creator function, by sending it an array of allowed origins—in this case, `http://localhost:3000`. This will make the API answer to the `OPTIONS` request with an `Access-Control-Allow-Origin: http://localhost:3000` header, which will signal to the browser that if the website making requests is running on `http://localhost:3000`, it should allow further requests.

This will work just fine. However, having this *hardcoded* domain here seems a little bit strange. We've been injecting all the similar configuration to the application. Remember what we did with the `port` configuration? Let's do the same for the allowed domains.

4. Change the `createServer` function parameters to receive an array of `string` named `allowedOrigins` inside `configuration` and later send it to the `oakCors` middleware creator function. The code for this is shown here:

```
interface CreateServerDependencies {
  configuration: {
    port: number,
    authorization: {
      key: string,
      algorithm: Algorithm
    },
    allowedOrigins: string[]
  },
  museum: MuseumController,
  user: UserController
}
…
```

```
export async function createServer({
  configuration: {
    port,
    authorization,
    allowedOrigins,
  },
  museum,
  user
}: CreateServerDependencies) {
  ...
app.use(
  oakCors({ origin: allowedOrigins })
);
```

We changed the type of the function parameters, used destructuring to get the variable from the arguments, and sent it into the oakCors middleware creator.

5. There's one thing missing, though—we need to send this array of allowedOrigins from src/index.ts. Let's do this, as follows:

```
createServer({
  configuration: {
    port: 8080,
    authorization: {
      key: authConfiguration.key,
      algorithm: authConfiguration.algorithm
    },
    allowedOrigins: ['http://localhost:3000']
  },
  museum: museumController,
  user: userController
})
```

And that should be all we need! We can now try to access it from a browser environment, from an application running at http://localhost:3000.

6. Let's test this, starting by running the API, as follows:

```
$ deno run --allow-net --unstable --allow-env --allow-
read --allow-write --allow-plugin src/index.ts
Application running at http://localhost:8080
```

7. To test it, create an HTML file named `index.html` in the root folder (museums-api), with a script that performs a POST request to `http://localhost:8080/api/users/register`. The code for this is shown here:

```
<!DOCTYPE html>
<html lang="en">
  <head>
    <meta charset="UTF-8" />
    <meta name="viewport" content="width=device-width,
       initial-scale=1.0" />
    <title>Test CORS</title>
  </head>
  <body>
    <div id="status"></div>
    <script type="module">
    fetch("http://localhost:8080/api/users/register", {
        body: JSON.stringify({ username: "abcd",
          password: "abcd" }),
        method: "POST",
    })
        .then(() => {
          document.getElementById("status").innerHTML
            = "WORKING";
        })
        .catch(() => {
          document.getElementById("status").innerHTML
            = "NOT WORKING";
        });
    </script>
  </body>
</html>
```

We're also adding a `div` tag and altering its inner HTML code in the cases that the request works or fails so that it's easier for us to diagnose.

In order for us to serve the HTML file and test this, you can leverage Deno and its ability to run remote scripts.

8. In the same folder we created the `index.html` file, let's run Deno's standard library web server, using the `-p` flag to set the port to `3000` and `--host` to set the host to `localhost`. The code for this is shown here:

```
$ deno run --allow-net --allow-read https://deno.land/
std@0.83.0/http/file_server.ts -p 3000 --host localhost
HTTP server listening on http://localhost:3000/
```

9. Access `http://localhost:3000` with your browser and you should see a **WORKING** message, as illustrated in the following screenshot:

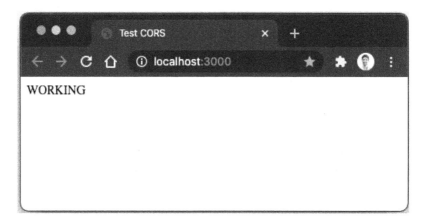

Figure 7.1 – Testing that the CORS API is working

10. If you want to test what happens when the origin is not in the `allowedOrigins` list, you can run the same command but with a different port (or host) and check the behavior. The code for this is shown here:

```
$ deno run --allow-net --allow-read https://deno.land/
std/http/file_server.ts -p 3001 --host localhost
HTTP server listening on http://localhost:3001/
```

You can now navigate to that new **Uniform Resource Locator** (**URL**) on the browser, and you should see a **NOT WORKING** message. If you look at the browser's console, you can also confirm that the browser is warning you that the CORS preflight request failed. That's the desired behavior.

And that's all we need to enable CORS on the API!

The third-party module we used has a few more options you can explore—things such as filtering for specific HTTP methods or answering the preflight request with a different status code. At the moment, the default options are working for us. We'll now proceed and see how we can enable users to connect to the application via HTTPS, adding an extra layer of security and encryption.

Enabling HTTPS

Any user-facing application nowadays should not only be allowing but also forcing its users to connect over HTTPS. This is a layer of security added on top of HTTP, making sure all connections are encrypted via a trusted certificate. Once again, we'll not try to come up with a definition, instead using the following one from MDN (`https://developer.mozilla.org/en-US/docs/Glossary/https`):

> *"HTTPS (HyperText Transfer Protocol Secure) is an encrypted version*
> *of the HTTP protocol. It uses SSL or TLS to encrypt all communication*
> *between a client and a server. This secure connection allows clients to safely*
> *exchange sensitive data with a server, such as when performing banking*
> *activities or online shopping."*

By enabling HTTPS connections in our application, we're making sure that it's way harder to intercept and interpret requests. Without this a malicious user can, for instance, intercept a login request and have access to the user's password-and-username combination. We're protecting the user's sensitive data.

As we're using `oak` in our application, we'll look for a solution on how to support HTTPS connections in its documentation. By looking at `https://doc.deno.land/https/deno.land/x/oak@v6.3.1/mod.ts`, we can see that the `Application.listen` method receives a `configuration` object, the same one we previously used to send the `port` variable. There are other options, though, as we can see here: `https://doc.deno.land/https/deno.land/x/oak@v6.3.1/mod.ts#Application`. That's what we'll use to enable HTTPS.

Let's see how we can change oak's configuration so that it supports secure connections, by following these steps:

1. Go to src/web/index.ts and add the secure, keyFile, and certFile options to the listen method call, as follows:

    ```
    await app.listen({
      port,
      secure: true,
      certFile: './certificate.pem',
      keyFile: './key.pem'
    });
    ```

    The certFile and keyFile properties expect a path to the certificate and the key files.

    If you don't have a certificate or you don't know how to create a self-signed one, no worries. Since this is only for learning purposes, you can use ours from the book's files at https://github.com/PacktPublishing/Deno-Web-Development/tree/master/Chapter07/sections/1-enabling-cors-and-https/museums-api. Here, you'll find certificate.pem and key.pem files that you can download and use. You can download them wherever you want in your computer, but we'll assume they're at the project root folder (museums-api) in the next code samples.

2. To keep our code tidy and more configurable, let's extract these options and send them as parameters to the createServer function, as follows:

    ```
    export async function createServer({
      configuration: {
        ...
        secure,
        keyFile,
        certFile,
      },
      ...
    }: CreateServerDependencies) {
    ```

3. This is what the CreateServerDependencies parameter type should look like:

    ```
    interface CreateServerDependencies {
      configuration: {
    ```

```
    port: number,
    authorization: {
      key: string,
      algorithm: Algorithm
    },
    allowedOrigins: string[],
    secure: boolean,
    keyFile: string,
    certFile: string
  },
  museum: MuseumController,
  user: UserController
}
```

4. And this is what the createServer function looks like afterward, with the destructured parameters:

```
export async function createServer({
  configuration: {
    port,
    authorization,
    allowedOrigins,
    secure,
    keyFile,
    certFile,
  },
  museum,
  user
}: CreateServerDependencies) {
  ...
  await app.listen({
    port,
    secure,
    keyFile,
    certFile
  });
```

5. To wrap up, we will now send the paths to the certificate and key files from the `src/index.ts` file, as follows:

```
createServer({
  configuration: {
    port: 8080,
    authorization: {
      key: authConfiguration.key,
      algorithm: authConfiguration.algorithm
    },
    allowedOrigins: ['http://localhost:3000'],
    secure: true,
    certFile: './certificate.pem',
    keyFile: './key.pem'
  },
  museum: museumController,
  user: userController
})
```

Now, to keep the logs accurate, we need to fix the event listener we previously created, which logs that the application is running. This handler should now take into consideration that the application might run over HTTP or HTTPS, and log according to that.

6. Go back to `src/web/index.ts` and fix the event listener that is listening for the `listen` event so that it checks whether the connection is secure or not. The code for this is shown here:

```
app.addEventListener('listen', e => {
  console.log(`Application running at
    ${e.secure ? 'https' : 'http'}://${e.hostname ||
      'localhost'}:${port}`)
})
```

7. Let's run the application and see if it works, as follows:

```
$ deno run --allow-net --unstable --allow-env --allow-
read --allow-plugin src/index.ts
Application running at https://localhost:8080
```

You should now be able to access that URL and connect to the application.

You might still be seeing security warnings, but no worries. You can click **Advanced** and **Proceed to localhost (unsafe)**, as illustrated in the following screenshot:

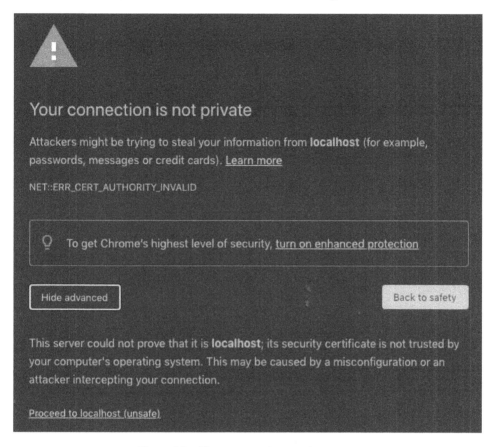

Figure 7.2 – Chrome security warning screen

This is due to the certificate being self-signed and not signed by a trusted certificate authority. However, it shouldn't matter much, as the process is the exact same as if it were a production certificate.

If you're still having problems, you might need to access the API URL directly before opening this page (`https://localhost:8080/`). From there, you can just follow the procedure on the following link (https://jasonmurray.org/posts/2021/thisisunsafe/) to enable communication with an API that isn't using a trusted certificate. After that, accessing `https://localhost:8080` will work.

The moment you have a proper certificate, signed by a trusted certificate authority, you can use it the same way we're using this one and everything will keep working just fine.

And that wraps it up for this section! We've added CORS and HTTPS support to our existing application, improving its security.

In the next section, we'll see how we can extract the configuration and secrets from our code, enabling it to be a little more flexible and configurable from the outside.

Let's go!

Extracting configuration and secrets

Any application, independent of its dimension, will have configuration parameters. By looking at the application we've been building in the previous chapters, even if we look at the simplest version of them all—the *Hello World* web server—we'll find configuration values, such as the `port` value.

It's also not a coincidence that we're sending a full object called `configuration` inside the `createServer` function, the function that starts up the web server. At the same time, we also have a couple of values that we know should be secret in the application. They're currently living in the code base, as it's been working for our purpose (which is learning), but we want to change it.

We're thinking of things such as the **JSON Web Token (JWT)** encryption keys, or the MongoDB credentials. Those are definitely not things you want to check out into your version control system. This is what this section is about.

We'll be looking at the configuration values and the secrets currently living in the code base. We will extract them so that they can be kept a secret and only passed to the application when it runs.

Doing this process can be a tough job when you have an application in which the configuration values are scattered across multiple modules and files. However, as we're following some architecture best practices and thinking about keeping the code decoupled and configurable, we made our life a little easier.

By having a look at `src/index.ts`, you can confirm that all the configuration values and secrets we're using are living there. This means that all the other modules are not aware of the configuration, and that's how it should be.

We'll be doing this "migration" in two phases. First, we'll extract all the configuration values into a `configuration` module, and then we'll extract the secrets.

Creating a configuration file

To start, let's find out what hardcoded values we have in the code that should be living in a configuration file. The following snippet highlights the values we don't want to have living in the code:

```
client.connectWithUri("mongodb+srv://deno-
  api:password@denocluster.wtit0.mongodb.net/
    ?retryWrites=true&w=majority")
const db = client.database("getting-started-with-deno");
...
const authConfiguration = {
  algorithm: 'HS512' as Algorithm,
  key: 'my-insecure-key',
  tokenExpirationInSeconds: 120
}

createServer({
  configuration: {
    port: 8080,
    authorization: {
      key: authConfiguration.key,
      algorithm: authConfiguration.algorithm
    },
    allowedOrigins: ['http://localhost:3000'],
    secure: true,
    certFile: './certificate.pem',
    keyFile: './key.pem'
  },
...
```

By looking at this snippet from our application code, we can already identify a few things, as follows:

- The cluster URL and database name (the username and password are secrets)
- JWT algorithm, and expiration time (the key is a secret)
- The web server port

- CORS allowed origins
- The HTTPS certificate and key file paths

These are the elements we're going to extract here. We'll start by creating what would be our configuration file with all these values.

We'll use **YAML Ain't Markup Language** (**YAML**), since this is a commonly used file type for configuration. If you're not familiar with it, no worries—it's quite simple to grasp. You can get an overview of how it works on the official website, `https://yaml.org/`.

We'll also make sure that we have different configuration files for different environments, thus creating a file that has the environment in its name.

Next, we'll implement a feature that will allow us to store our configurations in a file, starting by creating the file itself, as follows:

1. Create a `config.dev.yaml` file at the root of the project and add all the configurations there, like this:

```
web:
  port: 8080
cors:
  allowedOrigins:
    - http://localhost:3000
https:
  key: ./key.pem
  certificate: ./certificate.pem
jwt:
  algorithm: HS512
  expirationTime: 120
mongoDb:
  clusterURI: deno-cluster.wtit0.mongodb.net/
    ?retryWrites=true&w=majority
  database: getting-started-with-deno
```

We now need a way of loading this file into our application. We'll be creating a module named `config` inside the `src` folder for that.

To read the configuration file, we'll use the filesystem functions we learned in *Chapter 2, The Toolchain*, together with the `encoding` package from Deno's standard library.

2. Create a `src/config` folder with a file named `index.ts` inside.

    Here, we'll define and export a function named `load`. This function will be responsible for loading the configuration file. The code for this is shown here:

    ```
    export async function load() {

    }
    ```

3. Since we're using TypeScript, we'll define the type of what will be our configuration file, adding that as a return type for the `load` function. This should match the structure of the configuration file we previously created. The code for this is shown here:

    ```
    import type { Algorithm } from "../deps.ts";

    type Configuration = {
      web: {
        port: number
      },
      cors: {
        allowedOrigins: string[],
      },
      https: {
        key: string,
        certificate: string
      },
      jwt: {
        algorithm: Algorithm,
        expirationTime: number
      },
      mongoDb: {
        clusterURI: string,
        database: string
      },
    }

    export async function load(): Promise<Configuration> {
      ...
    ```

4. Inside the `load` function, we should now try to load the configuration file we previously created, by using Deno filesystem APIs. As there can be multiple files depending on the environment, we'll also add `env` as a parameter to the `load` function, with the default value of `dev`, as follows:

```
export async function load(env = 'dev'):
  Promise<Configuration> {
  const configuration = await Deno.readTextFile
    (`./config.${env}.yaml`) as Configuration;
...
```

With this, we have access to the contents of the configuration file. However, it's currently just a text file. We want to get it as a JavaScript `Object` so that we can access it. For this, we'll use the YAML encoding functionality from the standard library.

5. Install the YAML encoder module from the Deno standard library, using `deno cache` to make sure we update the `lock` file (refer to *Chapter 3*, *The Runtime and Standard Library*), and export it on `src/deps.ts`, as follows:

```
export { parse } from
  "https://deno.land/std@0.71.0/encoding/yaml.ts"
```

6. Import it on `src/config/index.ts` and use it to parse the contents of the read file, as follows:

```
import { Algorithm, parse } from "../deps.ts";
...
export async function load(env = 'dev'):
  Promise<Configuration> {
  const configuration = parse(await Deno.readTextFile
    (`./config.${env}.yaml`)) as Configuration;

  return configuration
}
```

We should now be ready to use the contents of the configuration in our application. Let's get back to `src/index.ts` and do it.

7. Import the `config` module, call its `load` function, and use the configuration values, which previously were hardcoded values.

    This is what the `src/index.ts` file should look like afterward:

    ```
    import { load as loadConfiguration } from
      './config/index.ts';
    const config = await loadConfiguration();
    ...
    client.connectWithUri(`mongodb+srv://
      deno-api:password @${config.mongoDb.clusterURI}`);
    ...
    const authConfiguration = {
      algorithm: config.jwt.algorithm,
      key: 'my-insecure-key',
      tokenExpirationInSeconds: config.jwt.expirationTime
    }
    ...
    createServer({
      configuration: {
        port: config.web.port,
        authorization: {
          key: authConfiguration.key,
          algorithm: authConfiguration.algorithm,
        },
        allowedOrigins: config.cors.allowedOrigins,
        secure: true,
        certFile: config.https.certificate,
        keyFile: config.https.key
      },
    ...
    ```

We should now be able to run our application as we did previously, with the difference that all our configuration is now living in a separate file.

And that's it regarding configuration! We've extracted the configurations from the code into a `config` file, making them easier to read and maintain. We've also created a module that abstracts all the configuration-file reading and parsing, making sure the rest of the application isn't concerned about that.

Next, we'll learn how we can extend this `config` module so that it also incorporates secret values read from the environment.

Accessing secret values

As I previously mentioned, we've used a couple of values that should be secret, but we initially kept them on the code. Those are values that might change from environment to environment, and configuration we want to keep as a secret for security reasons. This requirement makes it impossible to have them checked out into version control, and thus they have to live somewhere else.

One common practice to do this is to get these values from the environment, using environment variables. Deno provides an API that we'll use to read from environment variables. We'll be extending the `config` module so that it also includes secret values on its exported object of type `Configuration`.

Here are the values that are supposed to be secrets and that are still living in the code:

- MongoDB username

- MongoDB password

- JWT encryption key

Let's get them out of the code, and add them to the `configuration` object by following these steps:

1. In `src/config/index.ts`, add the MongoDB username and password to the configuration and the key to JWT in the configuration type, as follows:

    ```
    type Configuration = {
      web: {…};
      cors: {…};
      https: {…};
      jwt: {
        algorithm: Algorithm;
        expirationTime: number;
        key: string;
    ```

```
      };
  mongoDb: {
    clusterURI: string;
    database: string;
    username: string;
    password: string;
  };
}
```

Having these properties already on the configuration type, we now need
a way to add them there. Let's extend the load function so that it extends the
configuration object.

2. Extend the configuration object to include the username and password
    missing properties on mongoDb and key on jwt, as follows:

```
export async function load(env = 'dev'):
  Promise<Configuration> {
  const configuration = parse(await Deno.readTextFile
    (`./config.${env}.yaml`)) as Configuration;

  return {
    ...configuration,
    mongoDb: {
      ...configuration.mongoDb,
      username: 'deno-api',
      password: 'password'
    },
    jwt: {
      ...configuration.jwt,
      key: 'my-insecure-key'
    }
  };
}
```

The only thing still to do is to get these values from the environment instead of having them hardcoded here. We'll use Deno's API for that, in order to access the environment (`https://doc.deno.land/builtin/stable#Deno.env`).

3. Use `Deno.env.get` to get the variables from the environment. We should also set a default value in case the `env` variable is not present. The code is shown in the following snippet:

```
export async function load(env = 'dev'):
 Promise<Configuration> {
   const configuration = parse(await Deno.readTextFile
     (`./config.${env}.yaml`)) as Configuration;

   return {
     ...configuration,
     mongoDb: {
       ...configuration.mongoDb,
       username: Deno.env.get
         ('MONGODB_USERNAME') ||'deno-api',
       password: Deno.env.get
         ('MONGODB_PASSWORD') || 'password'
     },
     jwt: {
       ...configuration.jwt,
       key: Deno.env.get('JWT_KEY') || 'insecure-key'
     }
   }
 }
```

4. Let's get back to `src/index.ts` and use the secret values that we just added to the `configuration` object, as follows:

```
client.connectWithUri
(`mongodb+srv://${config.mongoDb.username}:
  ${config.mongoDb.password}
    @${config.mongoDb.clusterURI}`);
const db = client.database(config.mongoDb.database);
...
const authConfiguration = {
```

```
    algorithm: config.jwt.algorithm,
    key: config.jwt.key,
    tokenExpirationInSeconds: config.jwt.expirationTime
}
```

We're now in the right conditions to run our application, sending the secret values as environment variables. Keep in mind that for the application to have access to the environment, it needs the --allow-env permission. Let's try it.

Just make sure you add the username and password values you previously created. The code can be seen in the following snippet:

```
$ MONGODB_USERNAME=add-your-username MONGODB_PASSWORD=add-
your-password JWT_KEY=add-your-jwt-key deno run --allow-net
--unstable --allow-env --allow-read --allow-plugin src/index.ts
Application running at https://localhost:8080
```

Now, if we try to register and log in, we'll verify that everything is working. The application is connected to MongoDB and is retrieving the JWT token properly—the secrets are working!

Note for Windows users

In Windows systems, you can use the set command (https://docs.microsoft.com/en-us/windows-server/administration/windows-commands/set_1) to set environment variable. Windows doesn't support setting environment variables inline, and thus, you have to run these commands prior to running the API. Throughout the book, we'll use the *nix syntax, but you'll have to use the set command if you're using Windows, as the following code demonstrates.

Following are the set commands for Windows:

```
C:\Users\alexandre>set MONGODB_USERNAME=your-username
C:\Users\alexandre>set MONGODB_PASSWORD=your-password
C:\Users\alexandre>set JWT_KEY=jwt-key
```

We just managed to have all the configurations and secrets extracted from the code! This step made the configurations easier to read and maintain by writing them to a file, and made the secrets more secure by sending them via the environment to the application, instead of having them on the code base.

We're getting closer to an application that can easily be deployed and configured in different environments, something we'll do in *Chapter 9, Deploying a Deno Application*.

In the next section, we'll leverage Deno's capabilities to bundle code for the browser, creating a very simple JavaScript client that connects to the API. That client can then be used by frontend clients so that the HTTP connection is abstracted; it will also share code and types with the API code.

Get on board!

Running Deno code in the browser

One of the things we mentioned in the previous chapters and one that we've considered as one of Deno's selling points is its full compatibility with ECMAScript6. This makes it possible for Deno code to be compiled and run on the browser. This compilation is something made by Deno itself, and the bundler is included in the toolchain.

This feature enables a whole load of possibilities. A lot of them are due to the capacity for sharing code between the API and the client, and that's what we'll explore in this section.

We'll build a very simple JavaScript client to interact with the Museums API we just built. This client can then be used by any browser application that wants to connect to the API. We'll write that client in Deno and bundle it so that it can be used by a client, or even served by the application itself.

The client we'll write will be a very rudimentary HTTP client, thus we'll not focus much on the code. We're doing this to demonstrate how can we reuse code and types from Deno to generate code that runs on the browser. At the same time, we'll also explain some advantages of keeping a client and its API together.

Let's start by creating a new module in the application, which we'll call `client`, as follows:

1. Create a folder inside `src` named `client`, with a file named `index.ts` inside it.

2. Let's create an exported method, `getClient`, which should return an instance of our API client with three functions: `login`, `register`, and `getMuseums`. The code for this is shown in the following snippet:

```
interface Config {
  baseURL: string;
}
export function getClient(config: Config) {
  return {
```

```
      login: () => null,
      register: () => null,
      getMuseums: () => null,
    };
  }
```

Note how we're getting a `config` object that hosts `baseURL`.

3. Now, it's just a matter of implementing the HTTP logic to make requests to the API. We'll not do a step-by-step guide to implementing this as it is quite straightforward, but you can access the full client on the book files (`https://github. com/PacktPublishing/Deno-Web-Development/blob/master/ Chapter07/sections/3-deno-on-the-browser/museums-api/src/ client/index.ts`).

This is what the `register` method will look like:

```
import type { RegisterPayload, LoginPayload,
  UserDto  } from "../users/types.ts";

...

const headers = new Headers();
headers.set("content-type", "application/json");

...


register: ({ username, password }: RegisterPayload):
  Promise<UserDto> => {
  return fetch(
    `${config.baseURL}/api/users/register`,
    {
      body: JSON.stringify({ username, password }),
      method: "POST",
      headers,
    },
  ).then((r) => r.json());
},

...
```

Note how we're importing types from the `users` module, adding them to our application. This will make our functions much more readable, and it will later allow us to have type-checking and completion when writing tests using the TypeScript client. We're also creating an object with `content-type` headers that will be used in all the requests.

By creating an HTTP client, we handle things such as authentication automatically. In this specific case, our client can automatically save the token and send it in future requests after the user logs in.

This is what the `login` method would look like:

```
export function getClient(config: Config) {
  let token = "";

  ...

  return {

    ...

    login: (
        { username, password }: LoginPayload,
    ): Promise<{ user: UserDto; token: string }> => {
      return fetch(
        `${config.baseURL}/api/login`,
        {
          body: JSON.stringify({ username, password }),
          method: "POST",
          headers
        },
      ).then((response) => {
          const json = await response.json();
token = json.token;

return json;
        });
    },
```

It's currently setting the `token` variable that lives on the client instance. That token is later added to authenticated requests such as the `getMuseums` function, as demonstrated in the following snippet:

```
getMuseums: (): Promise<{ museums: Museum[] }> => {
  const authenticatedHeaders = new Headers();
  authenticatedHeaders.set("authorization", `Bearer
  ${token}`);
  return fetch(
    `${config.baseURL}/api/users/register`,
    {
      headers: authenticatedHeaders,
    },
  ).then((r) => r.json());
},
```

After creating the client, we want to distribute it. We can use the Deno bundle command to do it, as we learned in *Chapter 2, The Toolchain*.

If we want to have it served by our web server, we can also do this by adding a handler that serves the bundled content of our client file. It would look something like this:

```
apiRouter.get("/client.js", async (ctx) => {
  const {
    diagnostics,
    files,
  } = await Deno.emit(
    "./src/client/index.ts",
    { bundle: "esm" },
  );

  if (!diagnostics.length) {
    ctx.response.type = "application/javascript";
    ctx.response.body = files["deno:///bundle.js"];

    return;
  }
});
```

You might need to go back to your .vscode/settings.json file and enable the unstable property so that it recognizes we're using unstable APIs. This is demonstrated in the following snippet:

```
{
    ...
    "deno.unstable": true
}
```

Note how we're using the unstable Deno.emit API and setting the content-type as application/javascript.

We're then sending the file emitted by Deno (deno:///bundle.js) as the request body.

This way, if a client performs a GET request to /api/client.js, it will bundle and serve the content of the client we just wrote. The end result will be a bundled, browser-compatible JavaScript file that can then be used by an application.

To finish, we'll use this client in an HTML file that will authenticate and get the museums from the API. Proceed as follows:

1. Create an HTML file in the root of the project, named index-with-client. html, as illustrated in the following code snippet:

```html
<!DOCTYPE html>
<html lang="en">
  <head>
    <meta charset="UTF-8" />
    <meta name="viewport" content="width=device-width,
      initial-scale=1.0" />
    <title>Testing client</title>
  </head>
  <body>
  </body>
</html>
```

2. Create a script tag and import the script directly from the API URL, as follows:

```html
<script type="module">
  import { getClient } from
    "https://localhost:8080/api/client.js";
</script>
```

3. Now, it's just a matter of using the client we built. We'll use it to log in (with a user you previously created) and get a list of museums. The code is illustrated in the following snippet:

```
async function main() {
  const client = getClient
    ({ baseURL: "https://localhost:8080" });
  const username = window.prompt("Username");
  const password = window.prompt("Password");

  await client.login({ username, password });

  const { museums } = await client.getMuseums();
  museums.forEach((museum) => {
    const node = document.createElement("div");
    node.innerHTML = `${museum.name} -
      ${museum.description}`;
    document.body.appendChild(node);
  });
}
```

We'll use `window.prompt` to get the username and password when the user accesses the page, and then we'll log in with that data and use it to get museums. After this, we'll just add that to the **Document Object Model** (**DOM**), creating a list of museums.

4. Let's spin up the application again, as follows:

```
$ MONGODB_USERNAME=deno-api MONGODB_PASSWORD=your-
password deno run --allow-net --allow-env --unstable
--allow-read --allow-plugin --allow-write src/index.ts
Application running at https://localhost:8080
```

5. And then, serving the frontend application, this time adding the `-cert` and `--key` flags with paths to the respective files, to run the file server with HTTPS, as illustrated in the following snippet:

```
$ deno run --allow-net --allow-read https://deno.land/
std@0.83.0/http/file_server.ts -p 3000 --host localhost
--key key.pem --cert certificate.pem
HTTPS server listening on https://localhost:3000/
```

6. We can now access the web page at `https://localhost:3000/index-with-client.html`, fill in the username and password, and get a list of museums on the screen, as illustrated in the following screenshot:

The Louvre - The world's largest art museum and a historic monument in Paris, France.

Figure 7.3 – Web page with a JavaScript client getting data from the API

To log in in the previous step, you need to use a user you previously registered on the application. If you don't have one, you can create it using the following command:

```
$ curl -X POST -d'{"username": "your-username",
"password": "your-password" }' -H 'Content-Type:
application/json' https://localhost:8080/api/users/
register
```

Make sure to replace `your-username` with the desired username, and `your-password` with the desired password.

And with that, we've finished our section about using Deno on the browser!

What we just did can be further explored, unlocking great amounts of potential; this is just a quick example that applies to our use case. This practice makes it easier for any browser application to integrate with the application we just wrote. Instead of having to deal with HTTP logic, clients would just have to call methods and receive their responses. As we saw, this client can also handle topics such as authentication and cookies automatically.

This section explored one feature that Deno enables: compiling code for the browser.

We've applied it in the context of our application by creating an HTTP client that abstracts the user from the API. This feature can be used to do lots of things and is currently being used in writing frontend JavaScript code inside Deno.

As we explained in *Chapter 2, The Toolchain*, the only thing we have to take into consideration when writing code for the browser is not using functions from the Deno namespace. By following these restrictions, we can very easily write code in Deno using all its advantages and compile it to JavaScript for distribution.

This was just an introduction to a very promising feature. This feature, as with Deno, is still in its initial stages, and many great uses for it will be discovered by the community. Now that you're also aware of it, I'm sure you'll also come up with great ideas.

Summary

This was a chapter in which we focused a lot on practices that bring our application closer to a state that we can deploy into production. We started by exploring basic security practices, adding the CORS mechanism and HTTPS to the API. These two features, which are pretty much standard in any application, are a big security improvement on what we already had.

Also, thinking about deploying the application, we also abstracted the configuration and secrets from the code base. We started by creating an abstraction that would deal with it so that the configuration is not scattered, and modules just receive their configuration values without any awareness of how they're loaded. Then, we proceeded to using those values in our current code base, something that revealed itself to be quite easy. This step removed any configuration values from the code and moved them to a configuration file.

Once done with configuration, we used the same abstraction created to deal with secrets in an application. We implemented a feature that loads values from environment variables and adds them to the application configuration. Then, we used those secret values where they were needed, with things such as the MongoDB credentials, and token-signing keys.

We finished the chapter by exploring a possibility offered by Deno since its first days: bundling code for the browser. Applying this feature to our application context, we decided on writing a JavaScript HTTP client to connect to the API.

This step explored part of the potential of sharing code between the API and client, unlocking a world of possibilities. With this, we explored how we can use this bundling feature of Deno to compile a file at runtime and serve it to the user. Part of this feature's advantages will also be explored in the next chapter, where we'll write unit and integration tests for our application. Part of those tests will use the HTTP client created here, leveraging one big advantage of this practice: having the client and server in the same code base.

In the next chapter, we'll focus deeply on testing. We'll write tests for the logic we wrote in the rest of the book, starting with the business logic. We'll learn how we can improve the reliability of a code base by adding tests, and how the layers and architecture we created are crucial when it comes to writing them. The tests we'll write will go from unit to integration tests, and we'll explore the use cases where they apply. We'll see the value added by tests when it comes to writing new features and maintaining old ones. Along the way, we'll learn about some new Deno APIs.

The code isn't done until the tests are written, and thus we'll write them to conclude our API.

Let's go!

Section 3: Testing and Deploying

In this section, you will be creating meaningful integration and unit tests that enable the application to grow, and will also learn how to containerize and deploy Deno applications in the cloud.

This section contains the following chapters:

8
Testing – Unit and Integration

Code isn't created until the respective tests have been written. Since you're reading this chapter, I'll assume we can agree on that statement. However, you might wondering, why haven't we written any tests? Fair enough.

We chose not to do this because we believe it would make the content harder to absorb. Since we wanted to keep you focused on learning Deno while building an application, we decided not to do this. The second reason is that we truly wanted a full chapter focused on testing; that is, this one.

Testing is a very important part of the software life cycle. It can be used to save time, to clearly state requirements, or just because you want to feel confident in rewriting and refactoring later. Independent of the motivation, one thing is certain: you'll write tests. I also truly believe that testing plays a big role in software design. Code that is easy to test is likely easy to maintain.

Since we're great advocates of the importance of testing, we couldn't consider this a complete guide to Deno without learning about it.

In this chapter, we'll write different kinds of tests. We'll start with unit tests, which are very valuable tests for the developer and maintenance life cycle. Then, we'll move on to integration tests, where we'll run the app and perform a few requests on it. We'll finish by using the client we wrote in the previous chapter. We'll do all of this while adding tests to the application we previously built, going step by step, and making sure the code we previously wrote is working properly.

This chapter will also demonstrate how some of the architectural decisions we made at the beginning of this book will pay off. This will be an introduction to how we can write simple mocks and clean, focused tests using Deno and its toolchain.

In this chapter, we will cover the following topics:

- Writing your first test in Deno
- Writing an integration test
- Testing the web server
- Creating integration tests for the application
- Testing the API together with the client
- Benchmarking parts of the application

Let's get started!

Technical requirements

The code that will be used in this chapter can be found at `https://github.com/PacktPublishing/Deno-Web-Development/tree/master/Chapter08/sections`.

Writing your first test in Deno

Before we start writing our test, it's important to remember a few things. The most important of them is, why are we testing?

There might be multiple answers to this question, but most of them will gesture toward guaranteeing that the code is working. You might also say that you use them so that you have flexibility when it comes to refactoring, or that you value having short feedback cycles when it comes to implementation – we can agree to both of these. Since we didn't write a test before implementing these features, the latter doesn't apply too much to us.

We'll keep these objectives in mind throughout this chapter. In this section, we'll write our first test. We'll use the application we wrote in the previous chapters and add tests to it. We'll write two types of tests: integration and unit tests.

Integration tests will test how different components of the application interact. Unit tests test layers in isolation. If we think of it as a spectrum, unit tests are closer to the code, while integration tests are closer to the user. On the very end of the user side, there are also end-to-end tests. Those are the tests that test the application by emulating the user behavior, which we won't cover in this chapter.

Parts of the patterns we used when developing the actual application, such as dependency injection and inversion of control, are of great use when it comes to testing. Since we developed our code by injecting all its dependencies, now, it's just a matter of mocking those dependencies on tests. Remember: code that is easy to test is normally easy to maintain.

The first thing we'll do is write tests for the business logic. Currently, since our API is quite simple, it doesn't have much business logic. Most of it is living on `UserController`, since `MuseumController` is very simple. We'll start with the latter.

To write tests in Deno, we'll need to use the following:

- The Deno test runner (covered in *Chapter 2, The Toolchain*)
- The `test` method from the Deno namespace (`https://doc.deno.land/builtin/stable#Deno.test`)
- The assertion methods from the Deno standard library (`https://doc.deno.land/https/deno.land/std@0.83.0/testing/asserts.ts`)

These are all part of Deno, distributed and maintained by the core team. There are many other libraries that can be used in tests that you can find in the community. We'll use what's provided by default in Deno as it works just fine and allows us to write clear and readable tests.

Let's go and learn how we can define a test!

Defining a test

Deno provides an API to define tests. This API, `Deno.test` (`https://doc.deno.land/builtin/stable#Deno.test`), provides two different ways to define a test.

One of them is the one we showed in *Chapter 2, The Toolchain*, and consists of calling it with two arguments; that is, the test name and a test function. This can be seen in the following example:

```
Deno.test("my first test", () => {})
```

The other way we can do this is by calling the same API, this time sending an object as an argument. You can send the function and the name of the test, plus a few other options, to this object, as you can see in the following example:

```
Deno.test({
    name: "my-second-test",
    fn: () => {},
    only: false,
    sanitizeOps: true,
    sanitizeResources: true,
});
```

These flags behaviors are very well-explained in the documentation (https://doc.deno.land/builtin/stable#Deno.test), but here's a summary for you:

- only: Runs only the tests that have this set to true and makes the test suite fail, so this should only be used as a temporary measure.
- sanitizeOps: Makes the test fail if all the operations that started on Deno's core are not successful. This flag is true by default.
- sanitizeResources: Makes the test fail if there are still resources running after the test finishes (this can indicate memory leaks). This flag makes sure tests have to have a teardown phase where resources are stopped, and is true by default.

Now that we know about the APIs, let's go write our first test – a unit test for the MuseumController function.

A unit test for MuseumController

In this section, we'll be writing a very simple test that will cover only the functionality we wrote in MuseumController, and no more.

It lists all the museums in the application, though it's currently not doing much and is only working as a proxy for `MuseumRepository`. We can create the test file and logic for this simple functionality by following these steps:

1. Create the `src/museums/controller.test.ts` file.

    The test runner will automatically consider files that have `.test` in their name as test files, among other conventions, as explained in *Chapter 2, The Toolchain*.

2. Declare the first test with the help of `Deno.test` (`https://doc.deno.land/builtin/stable#Deno.test`):

    ```
    Deno.test("it lists all the museums", async () => {});
    ```

3. Now, export the assertion methods from the standard library under a namespace named `t`, so that we can then use them on the test files, by adding the following to `src/deps.ts`:

    ```
    export * as t from
      "https://deno.land/std@0.83.0/testing/asserts.ts";
    ```

    If you want to know what assertion methods are available in the standard library, check out `https://doc.deno.land/https/deno.land/std@0.83.0/testing/asserts.ts`.

4. You can now use the assertion methods from the standard library to write a test that instantiates `MuseumController` and calls the `getAll` method:

    ```
    import { t } from "../deps.ts";
    import { Controller } from "./controller.ts";

    Deno.test("it lists all the museums", async () => {
      const controller = new Controller({
        museumRepository: {
          getAll: async () => [{
            description: "amazing museum",
            id: "1",
            location: {
              lat: "123",
              lng: "321",
            },
            name: "museum",
    ```

```
      }],
    },
  });

  const [museum] = await controller.getAll();

  t.assertEquals(museum.name, "museum");
  t.assertEquals(museum.description, "amazing
    museum");
  t.assertEquals(museum.id, "1");
  t.assertEquals(museum.location.lat, "123");
  t.assertEquals(museum.location.lng, "321");
});
```

Note how we're instantiating MuseumController and sending in a mocked version of museumRepository, which returns a static array. This is how we're sure we're testing only the logic inside MuseumController, and nothing more. Closer to the end of the snippet, we're making sure the getAll method's result is returning the museum being returned by the mocked repository. We are doing this by using the assertion methods we exported from the dependencies file.

5. Let's run the test and verify that it's working:

```
$ deno test --unstable --allow-plugin --allow-env
--allow-read --allow-write --allow-net src/museums
running 1 tests
test it lists all the museums ... ok (1ms)


test result: ok. 1 passed; 0 failed; 0 ignored; 0
measured; 0 filtered out (1ms)
```

And our first test works!

Note how the test's output lists the name of the test, its status, and the time it took to run, together with a summary of the test run.

The logic inside `MuseumController` is quite simple, thus this was also a very simple test. However, it isolated the controller's behavior, allowing us to write a very focused test. If you're interested in creating unit tests for other parts of the application, they're available in this book's repository (`https://github.com/PacktPublishing/Deno-Web-Development/tree/master/Chapter08/sections/7-final-tested-version/museums-api`).

In the next few sections, we'll write more interesting tests. These are the tests that will teach us how to check the integration between the different modules of the application.

Writing an integration test

Our first unit test, which we created in the previous section, relied on a mocked instance of the repository to guarantee that our controller was working. That test adds great value when it comes to detecting bugs in `MuseumController`, but it isn't worth much in terms of understanding if the controller works well with the repository.

That's the purpose of integration tests: they test how multiple components integrate with each other.

In this section, we'll write the integration test that tests `MuseumController` and `MuseumRepository`. These are the tests that will closely mimic what happens when the application runs and will help us later in terms of detecting any problems between these two classes.

Let's get started:

1. Create the file for this module's integration tests inside `src/museums`, called `museums.test.ts`, and add the first test case there.

    It should test whether it is possible to get all the museums, this time using an instance of the repository instead of a mocked one:

    ```
    Deno.test("it is able to get all the museums from
        storage", async () => {});
    ```

2. We'll start by instantiating the repository and adding a couple of fixtures there:

    ```
    import { t } from "../deps.ts";
    import { Controller, Repository } from "./index.ts";

    Deno.test("it is able to get all the museums from
        storage", async () => {
    ```

```
const repository = new Repository();

repository.storage.set("0", {
  description: "museum with id 0",
  name: "my-museum",
  id: "0",
  location: { lat: "123", lng: "321" },
});

repository.storage.set("1", {
  description: "museum with id 1",
  name: "my-museum",
  id: "1",
  location: { lat: "123", lng: "321" },
});
...
```

3. Now that we have a repository, we can use it to instantiate the controller:

```
const controller = new Controller({ museumRepository:
  repository });
```

4. We can now write our assertions to make sure everything is working fine:

```
const allMuseums = await controller.getAll();

t.assertEquals(allMuseums.length, 2);

t.assertEquals(allMuseums[0].name, "my-museum", "has
  name");
t.assertEquals(
  allMuseums[0].description,
  "museum with id 0",
  "has description",
);
t.assertEquals(allMuseums[0].id, "0", "has id");
t.assertEquals(allMuseums[0].location.lat, "123", "has
  latitude");
```

```
    t.assertEquals(allMuseums[0].location.lng, "321", "has
        longitude");
```

Note how we're sending a message as a third argument to `assertEquals`, allowing us to get a proper message when this assertion fails. This is something that all assertion methods support.

5. Let's run the test and check the result:

```
$ deno test --unstable --allow-plugin --allow-env
--allow-read --allow-write --allow-net src/museums
running 2 tests
test it lists all the museums ... ok (1ms)
test it is able to get all the museums from storage ...
ok (1ms)


test result: ok. 2 passed; 0 failed; 0 ignored; 0
measured; 0 filtered out (2ms)
```

It is passing! This is all we need for our repository and controller integration tests! This test is useful whenever we want to change the code in `MuseumController` or `MuseumRepository` as it makes sure they work fine together.

Again, if you are curious about how integration tests for other parts of the application work, we made them available in this book's repository (`https://github.com/PacktPublishing/Deno-Web-Development/tree/master/Chapter08/sections/7-final-tested-version/museums-api`).

In the first section, we created a unit test, and here, we created an integration test, but we still haven't written any tests for our application's interface – the web part of it, which is using HTTP. That's what we'll do in the next section. We'll learn how can we test the logic living in the web layer in isolation, without using any other modules.

Testing the web server

So far, we have learned how to test different parts of the application. We started with the business logic, which tests how it integrated with the modules that interacted with persistency (the repository), but the web layer still has no tests.

It's true that those tests are very important, but we can agree that if the web layer fails, the user will not have access to any of that logic.

That's what we'll do in this section. We'll spin up our web server, mock its dependencies, and make a few requests to it to ensure the web *unit* is working.

Let's start by creating the web module's unit test by following these steps:

1. Go to `src/web` and create a file named `web.test.ts`.

2. Now, in order to test the web server, we need to go back to the `createServer` function in `src/web/index.ts` and export the `Application` object it creates in `src/web/index.ts`:

    ```
    const app = new Application();

    ...

    return { app };
    ```

3. We also want to be able to stop the application whenever we want. We haven't implemented that yet.

    If we look at oak's documentation, we'll see that it's very well-documented (`https://github.com/oakserver/oak#closing-the-server`).

    To abort the application that's started by the `listen` method, we also need to return `AbortController`. So, let's do that at the end of the `createServer` function.

    If you aren't aware of what `AbortController` is, I'll leave you with a link from Mozilla Developers Network (`https://developer.mozilla.org/en-US/docs/Web/API/AbortController`), which explains it very clearly. The short version is that it allows us to cancel an ongoing promise:

    ```
    const app = new Application();

    ...

    const controller = new AbortController();
    const { signal } = controller;

    ...

    return { app, controller };
    ```

    Note how we're instantiating `AbortController`, similar to the documentation's example, and returning it at the end, together with the `app` variable.

4. Back to our tests, let's create a test that checks whether the server answers to `hello world`:

    ```
    Deno.test("it responds to hello world", async () => {})
    ```

5. Let's get an instance of the server running using the function we previously created; that is, `createServer`. Remember, to call this function, we must send its dependencies in. Here, we'll have to mock them:

```
import { Controller as UserController } from
    "../users/index.ts";
import { Controller as MuseumController } from
    "../museums/index.ts";
import { createServer } from "./index.ts";
...
const server = await createServer({
  configuration: {
    allowedOrigins: [],
    authorization: {
      algorithm: "HS256",
      key: "abcd",
    },
    certFile: "",
    keyFile: "",
    port: 9001,
    secure: false,
  },
  museum: {} as MuseumController,
  user: {} as UserController,
});
```

We're sending in a configuration we'll be using for tests, using port `9001` and with HTTPS disabled, along with some random algorithm and key.

Note how we're using TypeScript's `as` keyword to pass mocked types into the `createServer` function without TypeScript warning us about the type.

6. We can now create a test that checks whether the web server is working by answering the hello world request:

```
import { t } from "../deps.ts";
...
const response = await fetch(
  "http://localhost:9001/",
  {
```

```
        method: "GET",
    },
).then((r) => r.text());

t.assertEquals(
    response,
    "Hello World!",
    "responds with hello world",
);
```

7. The last thing we need to do is close the server once the test has run. Deno makes the test fail by default if we don't do this (because `sanitizeResources` is `true` by default), as it would probably cause a memory leak:

```
server.controller.abort();
```

This wraps up our first test for the web layer! This was another unit test, and it tested the logic to spin up the server and ensured that Hello World is working. Next, we'll create more complete tests for the endpoints, together with the business logic.

In the next section, we'll start writing integration tests for the login and register functionality. Those are a little more complex than the tests we wrote for the museum's module as they'll test the application as a whole, including its business logic, persistency, and web logic.

Creating integration tests for the application

The three tests we've written so far have been unit tests for a single module, and an integration test between two different modules. However, to be confident that our code is working, it would be cool if we could test the application as a whole. That's what we'll do here. We'll wire up our application with a testing configuration and run a few tests against it.

We'll start by calling the same function we called to initialize the web server and then create instances of all its dependencies (controllers, repositories, and so on). We'll make sure we use things such as in-memory persistence to do so. This will make sure that our tests are replicable and don't need a complex teardown phase or a connection to a real database, as that would slow down the tests.

We'll start by creating a test file that, for now, will encompass the integration tests for the application. As the application evolves, it might make sense to create a test folder inside each module, but for now, this solution will work just fine.

We'll instantiate the application with a setup that's very close to what it runs in production and make a few requests and assertions against it:

1. Create the `src/index.test.ts` file, alongside the `src/index.ts` file. Inside it, create a test declaration that tests that a user can log in:

    ```
    Deno.test("it returns user and token when user logs
        in", async () => {})
    ```

2. Before we start writing this test, we'll create a helper function that will set up the web server for testing. It will contain all the logic for instantiating controllers and repositories, as well as sending configuration into the application. It will look something like this:

    ```
    import { CreateServerDependencies } from
        "./web/index.ts";

    ...

    function createTestServer(options?:
    CreateServerDependencies) {
      const museumRepository = new MuseumRepository();
      const museumController = new MuseumController({
        museumRepository });

      const authConfiguration = {
        algorithm: "HS256" as Algorithm,
        key: "abcd",
        tokenExpirationInSeconds: 120,
      };

      const userRepository = new UserRepository();
      const userController = new UserController(
        {
          userRepository,
          authRepository: new AuthRepository({
            configuration: authConfiguration,
          }),
    ```

```
      },
    );

    return createServer({
      configuration: {
        allowedOrigins: [],
        authorization: {
          algorithm: "HS256",
          key: "abcd",
        },
        certFile: "abcd",
        keyFile: "abcd",
        port: 9001,
        secure: false,
      },
      museum: museumController,
      user: userController,
      ...options,
    });
}
```

What we're doing here is very similar to the wiring logic we do in `src/index.ts`. The only difference is that we'll explicitly import the in-memory repositories, not the MongoDB ones, as shown in the following code block:

```
import {
  Controller as MuseumController,
  InMemoryRepository as MuseumRepository,
} from "./museums/index.ts";
import {
  Controller as UserController,
  InMemoryRepository as UserRepository,
} from "./users/index.ts";
```

For us to have access to the in-memory repositories of the Museums and Users modules, we need to go into these modules and export them.

This is what the `src/users/index.ts` file should look like:

```
export { Repository } from "./repository/mongoDb.ts";
export { Repository as InMemoryRepository } from
    "./repository/inMemory.ts";
export { Controller } from "./controller.ts";
```

This makes sure that we're exporting the default repository (which works with MongoDB) as `Repository` but also exporting `InMemoryRepository` at the same time.

Now that we have a way to create a test server instance, we can go back to writing our tests.

3. Create a server instance using the helper function we just created, `createTestServer`, and use `fetch` to make a register request to the API:

```
Deno.test("it returns user and token when user logs
    in", async () => {
    const jsonHeaders = new Headers();
    jsonHeaders.set("content-type", "application/json");
    const server = await createTestServer();

    // Registering a user
    const { user: registeredUser } = await fetch(
        "http://localhost:9001/api/users/register",
        {
            method: "POST",
            headers: jsonHeaders,
            body: JSON.stringify({
                username: "asantos00",
                password: "abcd",
            }),
        },
    ).then((r) => r.json())
    ...
```

4. Since we have access to the registered user, we can try to log in with that same user:

```
// Login in with the createdUser
const response = await
  fetch("http://localhost:9001/api/login", {
    method: "POST",
    headers: jsonHeaders,
    body: JSON.stringify({
      username: registeredUser.username,
      password: "abcd",
    }),
  }).then((r) => r.json())
```

5. We are now ready to develop a few assertions to check whether our login response is what we were expecting:

```
t.assertEquals(response.user.username, "asantos00",
  "returns username");
t.assert(!!response.user.createdAt, "has createdAt
  date");
t.assert(!!response.token, "has token");
```

6. Finally, we need to call the abort function on our server:

```
server.controller.abort();
```

This was our first application integration test! We got the application to run, performed the register and login requests against it, and asserted that everything was working as expected. Here, we built the test step by step, but if you want to have a look at the complete test, it is available in this book's GitHub repository (https://github.com/PacktPublishing/Deno-Web-Development/blob/master/Chapter08/sections/7-final-tested-version/museums-api/src/index.test.ts).

To wrap this up, we'll write another test. Remember that, in the previous chapter, we created some authorization logic that would only allow a logged in user to access the list of museums? Let's check if that is working with another test:

1. Create another test inside src/index.test.ts that will test whether a user with a valid token can access the museums list:

```
Deno.test("it should let users with a valid token
  access the museums list", async () => {})
```

2. Since we want to log in and register again, we'll extract those functions into a utility function that we can use in multiple tests:

```
function register(username: string, password: string) {
  const jsonHeaders = new Headers();
  jsonHeaders.set("content-type", "application/json");
  return
    fetch("http://localhost:9001/api/users/register", {
      method: "POST",
      headers: jsonHeaders,
      body: JSON.stringify({
        username,
        password,
      }),
    }).then((r) => r.json());
}

function login(username: string, password: string) {
  const jsonHeaders = new Headers();
  jsonHeaders.set("content-type", "application/json");
  return fetch("http://localhost:9001/api/login", {
    method: "POST",
    headers: jsonHeaders,
    body: JSON.stringify({
      username,
      password,
    }),
  }).then((r) => r.json());
}
```

3. With these functions, we can now refactor the previous test so that it looks a little cleaner, as the following snippet demonstrates:

```
Deno.test("it returns user and token when user logs
  in", async () => {
  const jsonHeaders = new Headers();
  jsonHeaders.set("content-type", "application/json");
  const server = await createTestServer();
```

```
// Registering a user
await register("test-user", "test-password");
const response = await login("test-user", "test-
password");

// Login with the created user
t.assertEquals(response.user.username, "test-user",
  "returns username");
t.assert(!!response.user.createdAt, "has createdAt
  date");
t.assert(!!response.token, "has token");

server.controller.abort();
});
```

4. Let's get back to the test we were writing – the one that checks whether an authenticated user can access the museums – and use the `register` and `login` functions to register and authenticate a user:

```
Deno.test("it should let users with a valid token
  access the museums list", async () => {
  const jsonHeaders = new Headers();
  jsonHeaders.set("content-type", "application/json");
  const server = await createTestServer();

  // Registering a user
  await register("test-user", "test-password");
  const { token } = await login("test-user", "test-
    password");
```

5. Now, we can use the token that's returned from the `login` function in the `Authorization` header to make an authenticated request:

```
const authenticatedHeaders = new Headers();
authenticatedHeaders.set("content-type",
  "application/json");
authenticatedHeaders.set("authorization",
  `Bearer ${token}`);
const { museums } = await
  fetch("http://localhost:9001/api/museums", {
    headers: authenticatedHeaders,
}).then((r) => {
  t.assertEquals(r.status, 200);

  return r;
}).then((r) => r.json());

t.assertEquals(museums.length, 0);

server.controller.abort();
});
```

Note that we're getting the token from the the `login` function and sending it with the `Authorization` header in the request to the museums route. Then, we're checking if the API responds correctly to the request with the 200 OK status code. In this case, since our application doesn't have any museums, it is returning an empty array, which we're also asserting.

Since we're testing this authorization feature, we can also test that a user with no token or an invalid token can't access this same route. Let's do it.

6. Create a test that checks that a user can't access the `museums` route without a valid token. It should be very similar to the previous test, with the small difference that we're sending an invalid token now:

```
Deno.test("it should respond with a 401 to a user with
  an invalid token", async () => {
  const server = await createTestServer();

  const authenticatedHeaders = new Headers();
  authenticatedHeaders.set("content-type",
    "application/json");
  authenticatedHeaders.set("authorization",
    `Bearer invalid-token`);
  const response = await
    fetch("http://localhost:9001/api/museums", {
      headers: authenticatedHeaders,
      body: JSON.stringify({
      username: "test-user",
      password: "test-password",
    }),
  });

  t.assertEquals(response.status, 401);
  t.assertEquals(await response.text(),
    "Authentication failed");

  server.controller.abort();
});
```

7. Now, we can run all the tests and confirm that they're all green:

```
$ deno test --unstable --allow-plugin --allow-env
--allow-read --allow-write --allow-net src/index.test.ts
running 3 tests
test it returns user and token when user logs in ...
Application running at http://localhost:9001
POST http://localhost:9001/api/users/register - 3ms
POST http://localhost:9001/api/login - 3ms
```

```
ok (24ms)
test it should let users with a valid token access
the museums list ... Application running at http://
localhost:9001
POST http://localhost:9001/api/users/register - 0ms
POST http://localhost:9001/api/login - 1ms
GET http://localhost:9001/api/museums - 8ms
ok (15ms)
test it should respond with a 400 to a user with an
invalid token ... Application running at http://
localhost:9001
An error occurred Authentication failed
ok (5ms)


test result: ok. 3 passed; 0 failed; 0 ignored; 0
measured; 0 filtered out (45ms)
```

This is it for the application integration tests we're going to write in this book! If you want to find out more, then don't worry – all the code that's been written in this book regarding tests is available in this book's GitHub repository (`https://github.com/PacktPublishing/Deno-Web-Development/tree/master/Chapter08/sections/7-final-tested-version/museums-api`).

We're now much more confident that our code is working. We've created the opportunity to refactor, extend, and maintain the code later with fewer worries. The architecture decisions we've made are paying off more and more when it comes to testing the code in isolation.

In the previous chapter, when we created our JavaScript client, we mentioned that one of the advantages of having it living in the API code base is that we could easily write tests for the client and the API to guarantee that they work well together. In the next section, we'll demonstrate how can we do this. These tests will be very much in line with what we did here, with the small difference that instead of using `fetch` and doing raw requests, we'll use the API client we created.

Testing the application together with the API client

When you provide an API client to your users, you have the responsibility of making sure it works flawlessly with your application. One of the ways to guarantee this is by having a complete test suite, one that not only tests the client on its own but also tests its integration with the API. Here we'll take care of the latter.

We'll use one feature of the API client and create a test that makes sure it's working. Once again, you'll notice some similarities between these and the tests we wrote at the end of the previous section. We'll replicate the logic from the previous tests, but this time we'll use the client. Let's get started:

1. Inside the same `src/index.test.ts` file, create a new test for the login functionality:

    ```
    Deno.test("it returns user and token when user logs in
      with the client", async () => {})
    ```

    For this test, we know that we'll need to get access to the API client. We'll need to import it from the `client` module.

2. Import the `getClient` function from `src/client/index.ts`:

    ```
    import { getClient } from "./client/index.ts"
    ```

3. Let's get back to the `src/index.test.ts` test and import `client`, thus creating an instance of it. Remember that it should use the same address that the test web server created:

    ```
    Deno.test("it returns user and token when user logs in
      with the client", async () => {
      const server = await createTestServer();

      const client = getClient({
        baseURL: "http://localhost:9001",
      });
      ...
    ```

    We can, of course, extract this server port to a variable that is used by both the `createTestServer` function and this test, but for simplicity, we won't do this here.

4. Now, it's just a matter of writing the logic that calls the `register` and `login` methods using `client`. This is what the final test will look like:

```
Deno.test("it returns user and token when user logs in
    with the client", async () => {
    ...
    // Register a user
    await client.register(
      { username: "test-user", password: "test-password"
        },
    );

    // Login with the createdUser
    const response = await client.login({
      username: "test-user",
      password: "test-password",
    });

    t.assertEquals(response.user.username, "test-user",
      "returns username");
    t.assert(!!response.user.createdAt, "has createdAt
      date");
    t.assert(!!response.token, "has token");
    ...
});
```

Note how we're using the client's methods to log in and register while keeping the assertions from the previous tests.

By following the same guidelines, we can write tests for all the client's functionality, guaranteeing that it's working fine with the API, making it easy to maintain it with confidence.

For brevity, and because these tests resemble the ones we previously written, we won't provide a step-by-step guide to writing tests for all the client's functionality here. However, if you're interested, you can find them in this book's GitHub repository (https://github.com/PacktPublishing/Deno-Web-Development/blob/master/Chapter08/sections/7-final-tested-version/museums-api/src/index.test.ts).

In the next section, we'll have a sneak peek at one feature that might be further down the path of your applications. On day, you'll start having parts of the application that seem to be getting slow and you want to track their performance, and that's where performance tests are useful. Because of this, we'll be introducing benchmarks.

Benchmarking parts of the application

When it comes to writing benchmarks in JavaScript, the language itself provides a few functions, all of which are included in the High Resolution Time API.

As Deno is fully ES6 compatible, these same features are available. If you've had the time to look at Deno's standard library or the official website, you'll have seen that benchmarks are taken into a lot of consideration and are tracked across Deno versions (`https://deno.land/benchmarks`). Upon checking Deno's source code, you will see that you have a very nice set of examples regarding how to write them.

For our application, we could easily use the APIs available on the browser, but Deno itself provides functionality in the standard library to help with writing and running benchmarks, so that's what we'll use here.

To start, we need to know Deno's standard library benchmark utilities so that we know what we can do (`https://github.com/denoland/deno/blob/ae86cbb551f7b88f83d73a447411f753485e49e2/std/testing/README.md#benching`). In this section, we'll write a very simple benchmark using two of the available functions; that is, `bench` and `runBenchmarks`. The first one will define a benchmark, while the second one will run it and print the result to the console.

Remember the function we wrote in *Chapter 5, Adding Users and Migrating to Oak*, to generate a hash and a salt, which enabled us to store the user credentials safely on the database? We'll write a benchmark test for that by following these steps:

1. To start, create a file alongside `src/users/util.ts` named `utilBenchmarks.ts`.

2. Import the two functions from `util` that we want to test; that is, `generateSalt` and `hashWithSalt`:

```
import { generateSalt, hashWithSalt } from "./util.ts"
```

3. It's time to add the benchmark utilities to our `src/deps.ts` file and run the `deno` `cache` command (which we learned about in *Chapter 2*, *The Toolchain*) and import it here. We'll export it as `benchmark`, in `src/deps.ts`, to avoid naming conflicts:

```
export * as benchmark from
    "https://deno.land/std@0.83.0/testing/bench.ts";
```

4. Import the benchmark utilities into our benchmarks file and write the first benchmark for the `generateSalt` function. We want it to run 1,000 times:

```
import { benchmarks } from "../deps.ts";

benchmarks.bench({
  name: "runsSaltFunction1000Times",
  runs: 1000,
  func: (b) => {
    b.start();
    generateSalt();
    b.stop();
  },
});
```

Note how we're sending an object to the `bench` function (as stated in the documentation). Inside this object, we're defining the number of runs, the name of the benchmark, and the test function. That function is what will run every time, since an argument is an object of the `BenchmarkTimer` type with two methods; that is, `start` and `stop`. These methods are used to start and stop the timings of the benchmarks, respectively.

5. The only thing we're missing is calling `runBenchmarks` once the benchmarks have been defined:

```
benchmarks.bench({
  name: "runsSaltFunction1000Times",
  ...
});

benchmarks.runBenchmarks();
```

6. It's time to run this file and have a look at the results.

   Remember that we're dealing with high resolution time as we want our benchmarks to be precise. To let this code have access to this system feature, we need to run this script with the `--allow-hrtime` permission (as explained in *Chapter 2, The Toolchain*):

```
$ deno run --unstable --allow-plugin --allow-env
--allow-read --allow-write --allow-hrtime src/users/
utilBenchmarks.ts
running 1 benchmarks ...
benchmark runsSaltFunction1000Times ...
    1000 runs avg: 0.036691561000000206ms
benchmark result: DONE. 1 measured; 0 filtered
```

7. Let's write the benchmark for the second function; that is, `hashWithSalt`:

```
benchmarks.bench({
  name: "runsHashFunction1000Times",
  runs: 1000,
  func: (b) => {
    b.start();
    hashWithSalt("password", "salt");
    b.stop();
  },
});


benchmarks.runBenchmarks();
```

8. Now, let's run it so that we get the final result:

```
$ deno run --allow-hrtime --unstable --allow-plugin
--allow-env --allow-write --allow-read src/users/
utilBenchmarks.ts
running 2 benchmarks ...
benchmark runsSaltFunction100Times ...
    1000 runs avg: 0.036691561000000206ms
benchmark runsHashFunction100Times ...
    1000 runs avg: 0.02896806399999923ms
benchmark result: DONE. 2 measured; 0 filtered
```

And that's it! You can now use the code we just wrote any time you want to analyze the performance of these functions. You may want to do this because you've changed this code or just because you want to have it closely tracked. You can integrate it in systems such as a continuous integration server, where you can regularly check these values and keep them on track.

This wraps up the benchmarks section of this book. We decided on giving it short introduction, and also demonstrating what APIs are available from on Deno to facilitate benchmarking needs. We believe the concepts and examples presented here will allow you to track how your applications are running.

Summary

With this chapter, we've closed the development cycle of the application we've been building. We started small by writing a few simple classes with our business logic, wrote the web server to it, and finished by integrating it with persistence. We finished this section by learning how to test the features we wrote, and that's what we did in this chapter. We decided on going with a few different types of tests, instead of extensively going module by module writing all the tests, as we believe that's where more value is added.

We started with a very simple unit test for the business logic, then moved on to an integration test with multiple classes, and later wrote a test for the web server. These tests can only be written by leveraging the architecture we've created, following dependency injection principles, and trying to keep the code as decoupled as possible.

As the chapter proceeded, we moved on to integration tests, which closely mimic the que application as it will run in production, enabling us to improve the confidence we have in the code we just wrote. We created tests that instantiated the application with a testing setup that enabled us to spin up the web server with all the application layers (business logic, persistence, and web) and made assertions to it. In these tests, we could very confidently assert that the login and registry behaviors were working fine, as we made real requests to the API.

To wrap up this chapter, we connected it to the previous one, where we wrote a JavaScript client for the API. We leveraged one of the big advantages of having the client living in the same codebase as the API and tested the client together with the application itself. This is a great way of guaranteeing that everything is working as expected, and that we can be confident when releasing changes in both the API and the client.

This chapter tried to demonstrate how tests can be used in Deno to increase our confidence in the code we've written, as well as the value they bring when they're used to focus on simple outcomes. Tests like these will be of great use later when the application changes, as we can use them to add more features or improve the existing ones. Here, we learned how the test suite provided by Deno is more than enough to write clear, readable tests without any third-party packages.

The next chapter will focus on one of the most important phases of application development; that is, deploying. We'll configure a very simple continuous integration environment where we can deploy the application to the cloud. This is a very important chapter as we'll also experience some of the advantages of Deno when it comes to its ease of deployment.

Excited to make your application available to users? So are we – let's go!

9
Deploying a Deno Application

Deployment is a crucial part of any application. We might build a great application, follow best practices, and write tests, but at the end of the day, when it gets to the user, this is where it will prove its value. As we want this book to be a journey through all the different phases of an application, we'll use this chapter about application deployment to close the cycle.

Note that we didn't—and will not—mention deployment as the final phase of software development, but as one phase of a cycle that will run multiple times. We truly believe that deployments shouldn't be events that everyone is afraid of. Rather, we see them as exciting moments whereby we're shipping features to our users. That's how most companies look at deployments in modern software projects, and we're true advocates of that. Deployments should be something regular, automated, and easy to do. They're the first step in getting features to our users, not the final step.

To enable this type of agility of processes and speed of iteration in applications, this chapter will focus on learning about containers and how to deploy a Deno application using them.

We'll take advantage of the benefits of containerization to create an isolated environment to install, run, and distribute our application.

As the chapter proceeds, we will learn how to use Docker together with `git` to create an automated workflow to deploy our Deno application in a cloud environment. Then, we'll tweak the way our application loads configurations to support having different configurations depending on the environment.

By the end of this chapter, we'll have the application running in a cloud environment and an automated process in place that enables us to ship iterations of it.

In this chapter, you'll get comfortable with the following topics:

- Preparing the environment for the application
- Creating a `Dockerfile` for the Deno application
- Building and running the application in Heroku
- Configuring the application for deployment

Technical requirements

The code used in this chapter can be found at the following GitHub link:

`https://github.com/PacktPublishing/Deno-Web-Development/tree/master/Chapter09`

Preparing the environment for the application

The environment where an application runs always has a big impact on it. It is one of the big causes of the so-common statement, *"it works on my machine"*. Over the years, developers have been creating solutions that try to minimize this as much as possible. These solutions can go from automatically provisioning new clean instances for the application to run, to creating more complete packages where everything the application depends on is included.

We can refer to **virtual machines** (**VMs**) or containers as ways to achieve this goal. Both are different solutions to the same problem but have one big thing in common: resource isolation. Both try to isolate an application from the environment around it. There are many reasons for this, from security, to automation, to reliability.

Containers are a modern way of providing a package for an application. Modern software projects use them to provide a single container image that has pretty much all it takes for an application to run.

If you're not aware of what a container is, I'll provide you with a definition from Docker's (a container engine) official website:

"A container is a standard unit of software that packages up code and all its dependencies so the application runs quickly and reliably from one computing environment to another."

In our path to make our application easily deployable, we will create this layer of isolation for our Deno application using Docker.

The end goal is to create an image that developers can use to deploy and test a specific version of the application. To do this with Docker, we need to configure the runtime where our application will run. This is defined in a file called a `Dockerfile`.

That's what we'll learn about next.

Creating a Dockerfile for the Deno application

A `Dockerfile` will allow us to specify what is required to create a new Docker image. This image will provide an environment containing all dependencies of the application, which can be used both for development purposes and for production deployments.

What we'll do in this section is learn how to create a Docker image for the Deno application. Docker provides a base image that is pretty much just the container runtime with isolation, called `alpine`. We could use that image, configure it, install all the tools and dependencies we need (namely Deno), and so on. However, I believe that we shouldn't be reinventing the wheel here, thus we're using a community Docker image.

Even though this image solves many of our problems, we still need to tweak it to our use case. Dockerfiles can be composed, which means they can extend other Docker images' functionality, and that's what we'll use.

> **Important note**
>
> As you might imagine, we'll not go deep in the fundamentals of Docker, as that would be a book in itself. If you're interested in Docker, you can start with the *Getting started* guide on the official documentation (`https://docs.docker.com/get-started/`). However, don't worry if you aren't currently very comfortable with Docker as we'll explain it enough for you to understand what we're doing here.

Before we start, make sure you install Docker Desktop on your machine by following the steps listed in the following link: https://docs.docker.com/get-docker/. After you have installed and started it, we have everything it takes to create our first Docker image! Let's create it by following these steps:

1. Create a Dockerfile at the root of our project.

2. As mentioned, we'll use an image from the community that already has Deno installed—hayd/deno (https://hub.docker.com/r/hayd/deno).

   This image is versioned in the same way as Deno, thus we'll use version 1.7.5. The FROM command from Docker allows us to extend an image, specifying its name and version tag, as illustrated in the following code snippet:

   ```
   FROM hayd/alpine-deno:1.7.5
   ```

3. The next thing we need to do is to define, inside the container, the folder we'll be working on.

   Docker containers provide a Linux filesystem, and the default workdir is the root of it (/). The WORKDIR command from Docker will allow us to work from a folder inside this same filesystem, making things a little bit tidier. The command can be seen here:

   ```
   WORKDIR /app
   ```

4. Now, we'll need to copy some files into our container image. With the help of the COPY command, we'll copy only the files we need for the installation step. In our case, these are the src/deps.ts and lock.json files, as illustrated in the following snippet:

   ```
   COPY lock.json .
   COPY src/deps.ts ./src/deps.ts
   ```

   The COPY command from Docker allows us to specify a file to copy from the local filesystem (the first parameter) into the container image (the last parameter), which is currently the app folder.

   By dividing our workflows and copying only the files we need, we allow Docker to cache and rerun part of the steps only when the involved files changed.

5. Having the files inside the container, we now need to install the application dependencies. We'll use deno cache to do this, as follows:

   ```
   RUN deno cache --lock=lock.json --unstable ./src/deps.ts
   ```

Since we're installing unstable dependencies (such as deno-mongo) and also using the lock file, we have to pass additional flags.

Docker's RUN command enables us to run this specific command inside the container.

6. Dependencies installed, we now need to copy the application's code into the container. One more time, we'll use Docker's COPY command for that, as shown here:

```
COPY . .
```

This will copy everything from the current local folder into the workdir (/app folder) inside the container.

7. The last thing we'll need to do for our image to work out of the box is introduce a command that will run whenever someone "executes" this image. We'll use Docker's CMD command to do this, as illustrated in the following snippet:

```
CMD ["deno", "run", "--allow-net", "--unstable",
"--allow-env", "--allow-read", "--allow-write", "--allow-
plugin", "src/index.ts" ]
```

This command takes an array of commands and parameters that will be executed when someone tries to run our image.

And that should be all we need to define our Deno application's Docker image! Having these capabilities in place will enable us to run our code locally in the same way it runs in production, a great advantage when it comes to debugging and investigating production issues.

The only thing we're missing is the actual step to generate the artifact.

We'll use the build command from the Docker **command-line interface** (**CLI**) in order to build the image. We'll use the -t flag to set the tag. Follow these steps to generate the artifact:

1. Inside the project folder, run the following command to generate the tag for the image:

```
$ docker build -t museums-api:0.0.1 .
```

You can use whatever name you want for the image (I have used museums-api in this example) and choose whichever version you want (0.0.1 in the example).

This should produce the following output:

```
$ docker build -t museums-api:0.0.1 .
Sending build context to Docker daemon   11.35MB
Step 1/7 : FROM hayd/alpine-deno:1.7.5
...
Step 2/7 : WORKDIR /app
...
Step 3/7 : COPY lock.json .
...
Step 4/7 : COPY src/deps.ts ./src/deps.ts
...
Step 5/7 : RUN deno cache --lock=lock.json --unstable ./
src/deps.ts
...
Step 6/7 : COPY . .
...
Step 7/7 : CMD ["deno", "run", "--allow-net",
"--unstable", "--allow-env", "--allow-read", "--allow-
write", "--allow-plugin", "src/index.ts" ]
...
Successfully built c3fa043b1440
Successfully tagged museums-api:0.0.1
```

And we have our image, museums-api:0.0.1. We could now publish it in a private image registry or use a public one such as Docker Hub. The **continuous integration** (**CI**) pipeline we'll set up later will be configured to perform this build step automatically.

What we can do now is run this image locally to verify that everything is working as expected.

2. To run the image locally, we'll use the Docker CLI run command.

    As we're dealing with a web application, we need to expose the port it is running on (set in the application's configuration file). We'll tell Docker to bind the container port to our machine's port by using the -p flag, as illustrated in the following code snippet:

```
$ docker run -p 8080:8080 museums-api:0.0.1
Download https://deno.land/std@0.83.0/encoding/hex.ts
Download https://deno.land/std@0.83.0/hash/mod.ts
```

```
Download https://deno.land/std@0.83.0/hash/_wasm/hash.ts
Download https://deno.land/std@0.83.0/hash/hasher.ts
Download https://deno.land/std@0.83.0/hash/_wasm/wasm.js
Download https://deno.land/std@0.83.0/encoding/hex.ts
Download https://deno.land/std@0.83.0/encoding/base64.ts
Check file:///app/src/index.ts
INFO downloading deno plugin "deno_mongo" from "https://
github.com/manyuanrong/deno_mongo/releases/download/
v0.13.0/libdeno_mongo.so"
INFO load deno plugin "deno_mongo" from local "/app/.
deno_plugins/deno_mongo_3bbff9a1cd76f3d988b3ca28c7163c3f.
so"
Application running at http://localhost:8080
```

This will execute version `0.0.1` of the `museums-api` image, binding the `8080` container port to the `8080` host port. We can now go to `http://localhost:8080` and see that the application is running.

We'll later use this image definition in a CI system that will create an image whenever the code changes and push it to production.

Having a Docker image containing an application can serve multiple purposes. One of them is this chapter's objective: deploying it; however, this same Docker image can also be used to run and debug an application at a specific version.

Let's learn how we can run a Terminal in a specific version of an application, a very common debug step.

Running a Terminal inside a container

Another useful thing we can do with a Docker image is execute a Terminal inside of it. This might be useful for debugging purposes or to try out something in a specific version of an application.

We can do that by using the same command as previously, together with a couple of different flags.

We'll use the `-it` flag, which will allow us to have an interactive connection to a Terminal inside the image. We'll also send, as a parameter, the name of the command we want to execute first inside the image. In this case it is `sh`, the standard Unix shell, as you can check in the following example:

```
$ docker run -p 8080:8080 -it  museums-api:0.0.1 sh
```

This will run the `museums-api:0.0.1` image, bind its `8080` port to the `8080` port on the host machine, and execute the `sh` command inside of it with an interactive Terminal, as illustrated in the following code snippet:

```
$ docker run -p 8080:8080 -it  museums-api:0.0.1 sh
/app # ls
Dockerfile              certificate.pem         config.staging.yaml
index.html              lock.json
README.md               config.dev.yaml         heroku.yml
key.pem                 src
```

Note that the folder where the shell is initially open is the one we've defined as our `WORKDIR` and that all our files are there. In the preceding example, we're also executing the `ls` command.

As we have an interactive shell attached to this container, we can use it to run a Deno command, for instance, as illustrated in the following code snippet:

```
/app # deno --version
deno 1.7.2 (release, x86_64-unknown-linux-gnu)
v8 8.9.255.3
typescript 4.1.3
/app #
```

This enables a full set of possibilities in terms of development and debugging, as we'll have the ability to check how the application is running in a specific version.

We've got to the end of this section. Here, we have explored containerization, introducing Docker and how it enables us to create an "application package". This package will take care of the environment around the application, making sure that it will run wherever there's a Docker runtime.

In the next section, we'll be using this same package to deploy an image we built locally in a cloud environment. Let's go!

Building and running the application in Heroku

As we mentioned when the chapter started, our initial objective was to have an easy, automated, and replicable way to deploy the application. In the previous section, we created our container image that will work as a basis for that. The next step is to create the pipeline that builds and deploys our code anytime there's an update. We'll use `git` as our source of truth and mechanism to trigger the pipeline builds.

The platform where we'll deploy our code is Heroku. This is a platform that aims to simplify tasks for developers and companies in the deployment process by providing a set of tools that removes common obstacles, such as provisioning machines and setting up big CI infrastructures. By using a platform such as this, we can be more focused on the application and on Deno, which is the purpose of this book.

Here, we'll use the `Dockerfile` that we previously created and set it up so that it is deployed and runs on Heroku. We'll see how easy it is to set up an application to run there, and later we'll also explore how we can define configuration values via environment variables.

Before we start, make sure you've created an account and installed the Heroku CLI before we proceed to the step-by-step guide, by following the two links provided here:

- Create an account: `https://signup.heroku.com/dc`.
- Install the CLI: `https://devcenter.heroku.com/articles/heroku-cli`.

Now that we have created an account and installed the CLI, we can start to set up our project in Heroku.

Creating the application in Heroku

Here, we will go through the steps needed to authenticate and create the application in Heroku. We're almost ready to start, but there's another thing we have to make clear first.

> **Important note**
> As Heroku uses `git` as source of truth, you'll *not* be able to do the following procedure inside the book's files repository, as it is already a Git repository containing multiple stages of the application.

What I recommend you do is copy the application files to a different folder, *outside the book's repository*, and start the process from there.

You can copy the latest version of the working application from *Chapter 8, Testing – Unit and Integration* (`https://github.com/PacktPublishing/Deno-Web-Development/tree/master/Chapter08/sections/7-final-tested-version/museums-api`), which is the one we'll use here.

Now that the files are copied into a new folder (outside the main repository), let's deploy the `Dockerfile` and run it on Heroku by following these steps:

1. The first thing we'll do is log in with the CLI, running `heroku login`. This should open a browser window where you can insert your username and password, as illustrated in the following code snippet:

    ```
    $ heroku login
    heroku: Press any key to open up the browser to login or
    q to exit:
    Opening browser to https://cli-auth.heroku.com/auth/
    cli/...
    Logging in... done
    Logged in as your-login-email@gmail.com
    ```

2. As Heroku deployments are based on `git`, and since we're now in a folder that is not a Git repository, we'll need to initialize it, as follows:

    ```
    $ git init
    Initialized empty Git repository in /Users/alexandre/dev/
    museums-api/.git/
    ```

3. Then, we will create our application in Heroku by using `heroku create`, as follows:

    ```
    $ heroku create
    Creating app... done, boiling-dusk-18477
    https://boiling-dusk-18477.herokuapp.com/ | https://git.
    heroku.com/boiling-dusk-18477.git
    ```

You might have noticed that when creating the application, the Heroku CLI automatically creates a Git repository in the current folder. Heroku automatically creates a remote repository called `heroku`, which is where we have to push our code to trigger the deployment process.

If you go to the Heroku Dashboard after running the preceding commands, you'll notice there's a new application living there. When the application is created, Heroku prints a URL on the console; however, as we haven't configured anything, our application is not available yet.

The next thing we'll need to do it to configure Heroku so that it knows it should build and execute our image in every deployment.

Building and running the Docker image

By default, Heroku just tries to make your application available by running the code. This is possible to do with many languages and you'll find guides for it in the Heroku documentation. As we want to use a container to run our application, the process needs a little more configuration.

Heroku provides a set of features that allow us to define what happens when there are changes in the code, via a file named `heroku.yml`. That's what we'll create now, as follows:

1. Create a `heroku.yml` file at the root of your repository and add the following lines of code there so that it builds our image using Docker, using the `Dockerfile` we created in the previous section:

    ```
    build:
      docker:
        web: Dockerfile
    ```

2. Now, in the same file, add the following lines of code to define the command that will be executed by Heroku to run the application:

    ```
    build:
      docker:
        web: Dockerfile
    run:
        web: deno run --allow-net --unstable --allow-env
    --allow-read --allow-write --allow-plugin src/index.ts
    ```

    You might notice that's the exact command we have defined inside the `Dockerfile`, and that's true.

Normally, Heroku would run the command from the `Dockerfile` to execute the image, and it would work. It happens that Heroku doesn't run these commands as root, as a security best practice. Deno, at its current stage, needs root privileges whenever you want to use plugins (an unstable feature).

As our application is using a plugin to connect with MongoDB, we need this command to be explicitly defined on `heroku.yml` so that it is run with root privileges and works when Deno is starting up the application.

3. The next thing we'll have to do is to set the application type to `container`, informing Heroku that's how we want this application to run. The code for this is shown in the following snippet:

```
$ heroku stack:set container
Setting stack to container... done
```

And that should be it for the Heroku configuration file. There are many other options that can be added there, as stated in the documentation (`https://devcenter.heroku.com/articles/build-docker-images-heroku-yml#heroku-yml-overview`), and used to do something more specific, but this works for our current use case.

The next thing we need to do is add all the files from this new repository (`heroku.yml` file included) to version control and push it to Heroku so that it starts the build.

4. Add all the files to make sure that `git` is tracking them:

```
$ git add .
```

5. Commit all the files with a message, as follows:

```
$ git commit -m "Configure heroku"
(root-commit) 9340446] Configure heroku
 34 files changed, 1465 insertions(+)
 create mode 100644
...
```

The `-m` flag that we've used is a command that allows us to create a commit with a message with a short syntax.

6. Now, it's a matter of pushing the files to the `heroku` remote.

This should trigger the build process of the Docker image, as you can check in the logs. This image is then pushed into Heroku's internal image registry in the last phase, as illustrated in the following code snippet:

```
$ git push heroku master
Enumerating objects: 42, done.
Counting objects: 100% (42/42), done.
Delta compression using up to 8 threads
Compressing objects: 100% (41/41), done.
Writing objects: 100% (42/42), 2.87 MiB | 1.22 MiB/s,
done.
Total 42 (delta 0), reused 0 (delta 0)
remote: Compressing source files... done.
remote: Building source:
remote: === Fetching app code
remote:
remote: === Building web (Dockerfile)
remote: Sending build context to Docker daemon   7.865MB
remote: Step 1/7 : FROM hayd/alpine-deno:1.7.2
remote: 1.3.0: Pulling from hayd/alpine-deno
...
```

Here, you can see that it's starting to build the `Dockerfile`, following all the steps specified there, as happened when we built the image locally, as illustrated in the following code snippet:

```
remote: === Pushing web (Dockerfile)
remote: Tagged image
"5c154f3fcb23f3c3c360e16e929c22b62847fcf8" as "registry.
heroku.com/boiling-dusk-18477/web"
remote: Using default tag: latest
remote: The push refers to repository [registry.heroku.
com/boiling-dusk-18477/web]
remote: 6f8894494a30: Preparing
remote: f9b9c806573a: Preparing
```

And it should be working, right? Well…, not really. We still have a couple of things that we need to configure, but we're almost there.

Remember that our application depends on the configuration, and that part of the configuration is coming from the environment. There's no way Heroku couldn't have known which configuration values we needed. There are still some settings we need to configure to get our application working, and that's what we'll do next.

Configuring the application for deployment

We now have an application that, when code is pushed to `git`, starts the process of building an image and deploying it. Our application currently gets deployed but it's not actually working, and this is happening because it is lacking configuration.

The first thing you probably noticed is that our application is always loading the configuration file from development, `config.dev.yml`, and it shouldn't.

When we first implemented this, we thought that different environments would have different configurations, and we were right. At the time, we didn't need to have configurations for more than one environment, and we used `dev` as a default. Let's fix that.

Remember that when we created the function that loads the configuration, we explicitly used an argument for the environment? We didn't use it at the time, but we left a default value.

Look at the following code snippet from `src/config/index.ts`:

```
export async function load(
    env = "dev",
): Promise<Configuration> {
```

What we'll need to do is change this so that it supports multiple environments. So, let's do that by following these steps:

1. Go back into `src/index.ts` and make sure we're sending the environment variable named DENO_ENV to the `load` function, as illustrated in the following code snippet:

    ```
    const config = await
        loadConfiguration(Deno.env.get("DENO_ENV"));
    ```

    This should keep the application running fine on development, where DENO_ENV is not defined, and allow us to load a different configuration file in production.

2. Create the production configuration file, `config.production.yml`.

    For now, it shouldn't be much different than `config.dev.yml`, with the exception of the `port`, for now. Let's get it running at port `9001` in production, as follows:

    ```
    web:
      port: 9001
    ```

    To test this locally, we can run the application with the `DENO_ENV` variable set to `production`, like this:

    ```
    $ DENO_ENV=production deno run --allow-net --unstable
    --allow-env --allow-read --allow-write --allow-plugin
    src/index.ts
    Application running at http://localhost:9001
    ```

    If you're running on Windows you will need to use a different syntax to set environment variables (like `DENO_ENV`). We mentioned how you can do this in *Chapter 7, HTTPS, Extracting Configuration, and Deno in the Browser*, in the *Accessing secret values* section.

    And after running it we can confirm it's loading the correct file, because the application port is now `9001`.

With what we just implemented, we're now able to control which configuration values are loaded based on the environment. This is something we already tested locally, but something we haven't done on Heroku.

We've solved part of the problem—we're loading a different configuration file depending on the environment, but there are other configuration values that our application depends on that are coming from the environment. Those are secret values such as the **JSON Web Token (JWT)** key or the MongoDB credentials.

There are many ways to do this, and all cloud providers provide a solution for that. In Heroku, we can do this by using the `config` command, as follows:

1. Define the MongoDB credential variables, the JWT key, and the environment using the `heroku config:set` command, as follows:

    ```
    $ heroku config:set MONGODB_PASSWORD=your-password
    MONGODB_USERNAME=your-username JWT_KEY=your-jwt-key DENO_
    ENV=production
    Setting MONGODB_PASSWORD, MONGODB_USERNAME, JWT_KEY,
    DENO_ENV and restarting boiling-dusk-18477... done, v7
    DENO_ENV:          production
    ```

```
JWT_KEY:             your-jwt-key
MONGODB_PASSWORD: your-password
MONGODB_USERNAME: your-username
```

Note how we're also defining the DENO_ENV variable so that our application knows that, when running in Heroku, it is the production environment.

If you are not using your own MongoDB cluster and you have questions about its credentials, you can go back to *Chapter 6, Adding Authentication and Connecting to the Database*, where we created a MongoDB cluster in MongoDB Atlas.

If you're using a different cluster, remember that it is defined in the configuration file in config.production.yml and not in the environment, and thus you need to add your cluster URL and database in the configuration file as follows:

```
...

mongoDb:
  clusterURI: <add your cluster url>
  database: <add your database name>
```

2. Once again, we'll add our changes to git, as follows:

    ```
    $ git commit -am "Configure environment variables and
    DENO_ENV"
    ```

3. And then, we'll push the changes to Heroku to trigger the deployment process, as follows:

    ```
    $ git push heroku master

    ...

    remote: Verifying deploy... done.
    To https://git.heroku.com/boiling-dusk-18477.git
       9340446..36a061e  master -> master
    ```

And it should be working. If we now go to the Heroku Dashboard (https://dashboard.heroku.com/), then into our application's dashboard (https://dashboard.heroku.com/apps/boiling-dusk-18477, in my case) and click the **Open Application** button, it should open our application, right?

Not yet, but we're almost there—we still need to sort out one more thing.

Getting the application port from the environment

Heroku has some particularities when it comes to running Docker images. It doesn't allow us to set the port where the application is running. What it does is assign a port where the application should run, and then redirect **HyperText Transfer Protocol** (**HTTP**) and **HyperText Transfer Protocol Secure** (**HTTPS**) traffic from the application URL there. If this still feels strange, no worries—we'll get there.

As you know, we explicitly defined the port our application was going to run on in the `config.production.yml` file. We need to adapt this.

The way Heroku defines which port an application should run on is by setting the `PORT` environment variable. This is documented at the following link:

`https://devcenter.heroku.com/articles/container-registry-and-runtime#dockerfile-commands-and-runtime`

You'll probably know from the title what we're doing next. We're going to change our application so that the web server port coming from the environment overrides the one defined in the configuration file.

Go back to `src/config/index.ts` in the application and make sure it is reading the `PORT` variable from the environment, overriding the configuration coming from the file. The code can be seen in the following snippet:

```
type Configuration = {
  web: {
    port: number;
  };
  cors: {
...
export async function load(
  env = "dev",
): Promise<Configuration> {
  const configuration = parse(
    await Deno.readTextFile(`./config.${env}.yaml`),
  ) as Configuration;

  return {
    ...configuration,
    web: {
      ...configuration.web,
```

```
    port: Number(Deno.env.get("PORT")) ||
        configuration.web.port,
    },
    ...
```

This way, we make sure that we're reading the variable from the PORT environment variable, using the value in the configuration file as a default.

And this should be all it takes to get our application running smoothly in Heroku!

Once again, we can test this by going to the Heroku Dashboard (`https://dashboard.heroku.com/apps/boiling-dusk-18477`) and clicking the **Open App** button, or you can do this by going directly to the URL—in my case, it's `https://boiling-dusk-18477.herokuapp.com/`.

Important note

If you are using MongoDB Atlas, as we did in *Chapter 6, Adding Authentication and Connecting to the Database*, and want to allow your application to access the database, you have to configure it so that it enables connections from "anywhere". This isn't recommended practice if you're exposing an application to your users, and it only happens because we're using Heroku's Free Tier. As it runs in a shared cluster, we have no way of knowing what the fixed **Internet Protocol** (**IP**) address is of the machine running the application, and we need to do it this way.

The following link demonstrates how can you configure network access to the database: `https://docs.atlas.mongodb.com/security/ip-access-list`. Make sure you click **ALLOW ACCESS FROM ANYWHERE** in the MongoDB Atlas network access screen.

This is what the network access screen looks like:

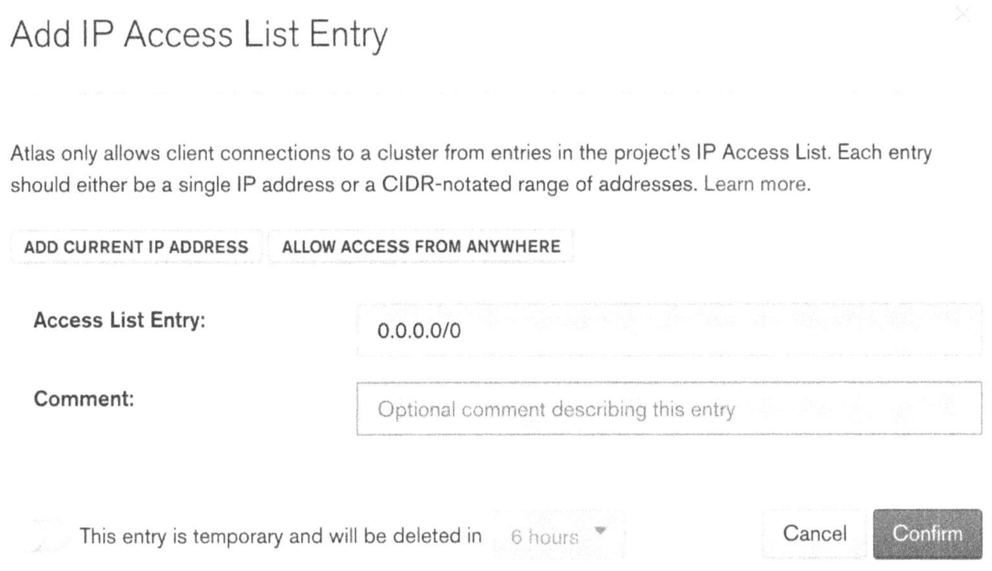

Add IP Access List Entry

Atlas only allows client connections to a cluster from entries in the project's IP Access List. Each entry should either be a single IP address or a CIDR-notated range of addresses. Learn more.

ADD CURRENT IP ADDRESS   ALLOW ACCESS FROM ANYWHERE

Access List Entry: 0.0.0.0/0

Comment: Optional comment describing this entry

This entry is temporary and will be deleted in   6 hours ▾   Cancel   Confirm

Figure 9.1 – MongoDB Atlas network access screen

After this, our application should work as expected; you can try to perform a request to register a user (that connects to the database) and check that everything's fine, as demonstrated in the following code snippet:

```
$ curl -X POST -d '{"username": "test-username-001",
"password": "testpwl" }' -H 'Content-Type: application/json'
https://boiling-dusk-18477.herokuapp.com/api/users/register
{"user":{"username":"test-username-001","createdAt":"2020-12-
19T16:49:51.809Z"}}%
```

If you got a response that's similar to the preceding one, you're all set! We managed to configure and deploy our application in a cloud environment and created an automated way to ship updates to our users.

To do a final test to check whether the code is being deployed successfully, we can try to change part of the code and trigger the deployment process again. Let's do it! Proceed as follows:

1. Change our "Hello World" message in src/web/index.ts to "Hello Deno World!", as illustrated in the following code snippet:

```
app.use((ctx) => {
  ctx.response.body = "Hello Deno World!";
});
```

2. Add this change to version control, as follows:

```
$ git commit -am "Change Hello World message"
[master 35f7db7] Change Hello World message
 1 file changed, 1 insertion(+), 1 deletion(-)
```

3. Push it into Heroku's git remote repository, like this:

```
$ git push heroku master
Enumerating objects: 9, done.
Counting objects: 100% (9/9), done.
Delta compression using up to 8 threads
Compressing objects: 100% (5/5), done.
Writing objects: 100% (5/5), 807 bytes | 807.00 KiB/s,
done.
Total 5 (delta 4), reused 0 (delta 0)
remote: Compressing source files… Done
…
remote: Verifying deploy... done.
To https://git.heroku.com/boiling-dusk-18477.git
```

4. If we now access the application URL (which is https://boiling-dusk-18477.herokuapp.com/, in our case), you should get the Hello Deno World message.

This means that our application was successfully deployed. Since we're using a cloud platform that provides more than what we learned here, we can explore other Heroku features, such as logging.

Next to the **Open app** button on the Heroku Dashboard (`https://dashboard.`
`heroku.com/`), you have a **More** button. One of the options is **View logs**, as you can see
in the following screenshot:

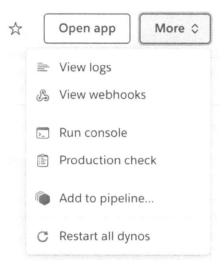

Figure 9.2 – More application options in the Heroku Dashboard

If you click there, an interface that shows the logs in real time will appear. You can try it
out by opening your application in a different tab (by clicking the **Open app** button).

You'll see that the logs instantly update, and something like this should appear there:

```
2020-12-19T17:04:23.639359+00:00 app[web.1]: GET http://
boiling-dusk-18477.herokuapp.com/ - 1ms
```

This is of great use when you want to have a very light monitoring of how your application
is running. The Logging feature is provided in the Free Tier, but there are many more
features you can explore, such as the **Metrics** one, which we'll not do here.

If you want to have a detailed look at when and by whom your application was deployed,
you can also use the **Activity** section from the Heroku Dashboard, as illustrated in the
following screenshot:

Figure 9.3 – Heroku Dashboard application options

You'll then see a log of your most recent deployments, another very interesting feature of Heroku, as illustrated in the following screenshot:

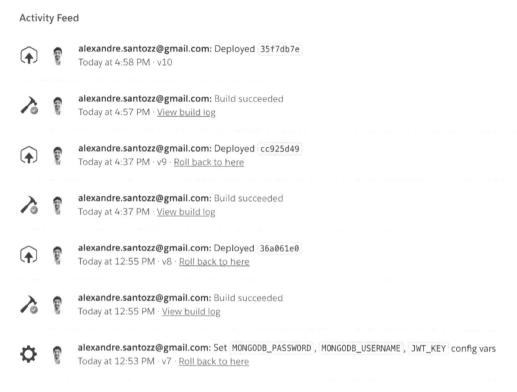

Figure 9.4 – Activity tab in the Heroku Dashboard application

This wraps up our section on deploying the application in a cloud environment.

We focused on the application and on topics that can and will be reused independent of the platform on which your code is running. We've iterated the application logic that loaded the configuration so that it could load different configurations depending on the environment.

Then, we learned how we could send environment variables with secret configuration values to our application, and we wrapped up by exploring logging on the platform of choice for this example, Heroku—and that was it.

We managed to get our application to run, and we created a whole infrastructure around it that will enable future iterations to be easily shipped to our users. Hopefully, we went through some of the phases you'll also go through next time you decide to deploy a Deno application.

Summary

And we're pretty much done! This chapter completes the cycle of development phases in our application by deploying it. We went from building a very simple application, to adding features to it, to adding tests, and—finally—to deploying it.

Here, we learned how we can use some of the benefits of containerization in our applications. We started learning about Docker, our container runtime of choice, and rapidly moved on to creating an image for our application. Learning about some Docker commands as we moved through the process, we also experienced how easy it is to prepare a Deno application to be deployed.

Creating this Docker image enabled us to have a replicable way of installing, running, and distributing our application, creating a package with everything the applications needs.

As the chapter proceeded, we started to explore how we can use this application package to deploy it in a cloud environment. We started by configuring the cloud platform of choice for this step-by-step guide, Heroku, so that it would rebuild and run our application's code every time it changed, and we very easily made it with the help of `git` and Heroku's documentation.

As the automated pipeline was configured, we understood the need to send configuration values to our application. These same configuration values, which we previously implemented in earlier chapters, needed to be sent to the application in two different ways, via a configuration file and via the environment. We tackled each one of those needs, first by iterating the application code so that it loaded different configurations depending on the environment, and later by learning how to set configuration values in an application living on Heroku.

We finally got our application to run flawlessly and completed the objective of the chapter: to have a replicable, automated way to deploy code to our users. In the meantime, we learned a bit about Docker and the benefits of containerization and automation when it comes to releasing code.

This pretty much wraps it up on the book's objective. We decided to make this a journey of building an application, separately going through all its phases and addressing them as needed. This was the last phase—deploying it, which hopefully closes the cycle for you in terms of going from the first line of code to deployment.

The next chapter will focus on what's next when it comes to Deno, both for the runtime and for you personally. I hope this has made you a Deno enthusiast and that you are as excited as I am about it and the world of possibilities it opens up.

10
What's Next?

We've come a long way. We started by getting to know the basics of Deno and went on to building and deploying a complete application. By now, you should be comfortable with Deno and have a good awareness of the problems it solves. Hopefully, all the phases we've been through have helped clarify many questions you may have had about Deno.

We deliberately chose to make this book a journey, which started with our first scripts and finished with a deployed application, one that we've coded and iterated as the book has proceeded. In the meantime, we solved many of the challenges an application developer might encounter and came up with solutions.

By now, you should be armed with the knowledge that will help you decide whether Deno will be part of the solution for your next project.

This chapter will begin with a short recap of what we have learned so far, going through all the phases and learnings. Then, our focus will go—as the title of the chapter suggests—toward the future. It will focus on what's next, both for Deno as a runtime and also for you as a developer with a new tool under your belt.

We'll have a quick look at what the current priorities of Deno's core team are, what they're working on, and what the proposed future features are. As the chapter proceeds, we'll also have a look at what's happening in the community, highlighting a few interesting initiatives.

The chapter will wrap up by demonstrating how we can publish a package to Deno's official registry, among other ways to give back to the Deno community.

By the end of this chapter, you'll be comfortable with the following areas:

- Looking back at our journey
- Deno's roadmap
- Deno's future and community
- Publishing a package to Deno's official registry

Looking back at our journey

A lot of ground was covered. We trust the book was (hopefully) an interesting journey, from not knowing Deno to building something with it, and finishing up with a deployed application.

We started by getting to know the tool itself, first by learning about the functionalities it provides, and then by writing simple programs with the standard library. As our knowledge built up we rapidly started to have enough to build a real application with it, and that's what we did.

The adventure began by building the simplest possible web server, using the standard library. We used TypeScript extensively to help in specifying clear application boundaries, and we managed to get a very simple application running, hitting our very first checkpoint: **hello world**.

Our application evolved, and as it started to have more complex requirements, we needed to dive into the web frameworks available on the Deno community. After making a high-level comparison between all of them, and according to our application needs, we went with oak. The next step was to migrate our (still) simple web server to use a framework of our choice, and it was a breeze. Using a web framework made our code simpler and enabled us to delegate things we really didn't want to handle ourselves, allowing us to focus on the application itself.

The next step was to add users to our application. We created the application endpoints to enable a user registry, and as the need to store users arose, we connected the application to MongoDB. With users in place, it was a short step to implement user authentication.

As the application grew, so did the need to have more complex configurations. From the server port it was running on to the location of certificate files, or to database credentials, all of this needed to be handled on its own. We abstracted the configuration from the application and centralized it. Along the way, we added support for configuration to live in a file or in environmental variables. This made it possible to run the application with different configurations depending on the environment, while keeping sensitive values safe and outside of the code base.

As our journey was coming to an end, we wanted to make sure our code was reliable enough. This pointed us toward a testing chapter, where we learned the basics of testing in Deno and created different tests for a few use cases of the application we had created. We went from a simple unit test to a cross-module test, to a test that got the application running, and made a few requests to it. By the end of the process, we had much more confidence that our code was working as expected, and we added testing capabilities to our toolchain.

To wrap things up, we turned the code we wrote into reality and we deployed it.

We got the application running on Heroku under a containerized environment. In the meantime, we learned about Docker and how it can be used to make it easy for developers to run and deploy their code. We finished this chapter with an automated way of deploying a Deno application, closing the cycle from code to deployment.

This was a journey whereby we went through the many common phases of an application's development, encountering challenges and solving them with solutions adapted to our use cases. I hope to have covered some of your main concerns and questions, giving you a sound basis to help you in the future.

We don't know what's next, but we do know that it depends on Deno and its community, and we hope that you see yourself as part of this. In the next section, we'll have a look at Deno's future roadmap, what's planned, and where their short-term efforts are directed.

Deno's roadmap

A lot has changed since the first time Ryan presented Deno on JSConf; a few big steps have been taken. With the first stable version of the runtime being launched the community exploded, and many people from other JavaScript communities joined in with many enthusiastic ideas.

Deno's core team is currently putting much of its efforts into pushing Deno forward. This contribution not only happens in the form of code, issues, and helping people, but also in planning and delineating what the next steps are.

For the short-term roadmap, the core team makes sure that it is tracking initiatives. The following two issues raised on GitHub have been used to track 2020's Q4 and 2021's Q1 efforts:

- `https://github.com/denoland/deno/issues/7915`
- `https://github.com/denoland/deno/issues/8824`

If you have a detailed look at these, you can follow every discussion, code, and decision that has been made regarding those features. I'll list some of the current initiatives here so that you can have a sneak peak of what's happening:

- The Deno **Language Server Protocol** (**LSP**) and language server
- Compilation to binary (single executable for a Deno application)
- Data, blob, WebAssembly, and **JavaScript Object Notation** (**JSON**) imports
- Improved support for Web Crypto **application programming interfaces** (**APIs**)
- Support for **Immediately Invoked Function Expressions** (**IIFE**) bundles
- WebGPU support
- HTTP/2 support

These are just a few examples of some of the important initiatives happening with Deno. As you can imagine, and due to it being in the early stages, there are currently a lot of efforts aimed at fixing bugs and refactoring code that I haven't added to this list.

Feel free to dive deeper into the GitHub issues mentioned previously to get more details about any of the initiatives.

All these are Deno's core team efforts. Remember that Deno only exists because there are people working on it in their free time. There are many ways to give back to the community, be that with bug reports, code contributions, helping on communication channels, or with donations.

If Deno is helping you and your company turn ideas into reality, please consider becoming a sponsor so that it stays healthy and keeps evolving. You can do this on GitHub at the following link: https://github.com/sponsors/denoland.

There are other people who are also responsible for Deno, the enthusiasm around it, and its evolution, and those people are Deno's community. In the next section, we'll go over Deno's community, interesting things happening there, and how can you play an active part in it.

Deno's future and community

The Deno community is growing rapidly—it is full of people who are excited about it and eager to help it grow. As you start using it, as you did throughout the course of this book, there will be very important contributions you can add to it. This could be a bug you've encountered, a feature that makes sense to you, or just something that you want to understand better.

For you to be part of that, I'd recommend you joining Deno's Discord channel (`https://discord.gg/deno`). This is a very active place where you can find other people interested in Deno and is useful if you want to find package authors, build packages yourself, or help with Deno Core. From my experience, I can only say that everyone I have met there is very friendly and eager to help. It's also a great way of keeping updated on what's happening.

Another way to contribute is by following Deno's repositories on GitHub (`https://github.com/denoland`). The main repository can be found at `https://github.com/denoland/deno`, where you'll find the Deno **command-line interface** (**CLI**) and Deno Core, while the standard library lives in its own repository (`https://github.com/denoland/deno_std`). There are also other repositories such as `https://github.com/denoland/rusty_v8`, which hosts the Rust bindings used by Deno created for the V8 JavaScript engine, or `https://github.com/denoland/deno_lint`, where the Deno linter is hosted, among others. Feel free to watch the repositories that interest you on GitHub.

> Tip
> A great way of being updated about what's happening on Deno without getting too many notifications is by watching Deno's main repository for releases only. You'll get a notification for every release whereby you can follow the very comprehensive release notes. I'll leave you with an example of a release note so that you know what they look like.

This is what a version update notification looks like:

Latest release

♢ v1.6.2

⊶ d199e45

Verified

Compare ▾

v1.6.2

🐙 **github-actions** released this 6 days ago · 10 commits to master since this release

1.6.2 / 2020.12.22

- feat(lsp): support the unstable setting (#8851)
- feat(unstable): record raw coverage into a directory (#8642)
- feat(unstable): support in memory certificate data for Deno.createHttpClient (#8739)
- fix: atomically write files to $DENO_DIR (#8822)
- fix: implement ReadableStream fetch body handling (#8855)
- fix: make DNS resolution async (#8743)
- fix: make dynamic import errors catchable (#8750)
- fix: respect enable flag for requests in lsp (#8850)
- refactor: rename runtime/rt to runtime/js (#8806)
- refactor: rewrite lsp to be async (#8727)
- refactor: rewrite ops to use ResourceTable2 (#8512)
- refactor: optimise static assets in lsp (#8771)
- upgrade TypeScript to 4.1.3 (#8785)

Changes in std version 0.82.0:

- feat(std/node): Added os.type (#8591)

Figure 10.1 – Deno's v1.6.2 release notes

On top of the GitHub releases shown in the preceding screenshot, the Deno team has also made efforts to write comprehensive release notes on their website, another great way of staying in the loop (https://deno.land/posts).

And this is what you can do to be an integral part of Deno's community. All it takes for you to start is for you to use it, report bugs, and meet new people, and the rest will follow.

The community is not only made of the core and the people who help with Deno, but also the packages and projects that have been built with it.

In the following section, I'll highlight some initiatives that I think are great and that are pushing the community forward. This is a personal list; take it as a recommendation and nothing more, as I'm sure there are other initiatives that could also be added.

Interesting things happening in the community

During the last two years that I've been following Deno, a lot of things have happened. After the v1.0.0 release and with more people joining, lots of ideas arose. I'll list a few initiatives that I think are especially interesting, not only for the functionality they provide but also as a great source of learning.

Denon

As Nodemon was the go-to solution when developing Node, Denon is one of the most used tools when it comes to Deno. If you haven't heard of it, it basically watches your files and reruns your Deno applications as soon as you change anything. It's one of those tools you most likely want to keep on your toolchain while developing with Deno. You can check out their GitHub page at `https://github.com/denosaurs/denon`.

Aleph.js

Even though we don't have space to explore it much here, Deno's capabilities to run on the browser unlock a full set of new functionalities, which has led to initiatives such as Aleph.js. This initiative calls itself the *React framework in Deno*, and it's been getting quite some usage and enthusiasm. If you haven't heard about it, it takes many aspects from the Next.js framework (`https://nextjs.org/`), implements them in Deno, and adds a few more things. It is quite new but already has features such as server-side rendering, hot module reloading, and filesystem and API routing, among others. You can read more about it at `https://alephjs.org/`.

Nest.land

Even though Deno has its own registry (which we'll use in the next section), there are still reasons why the community created other registries. Nest.land is one of them; it is a module registry based on blockchain technology that makes sure modules hosted there are not deleted. It is free, decentralized, and works without Git, and is the go-to solution for many package authors. Read more about it at `https://nest.land/`.

Pagic

As static site generators continue to get more and more usage, it was just a matter of time until some of them were made with Deno. That's what Pagic does—it's a static site generator with interesting features such as React, Vue, and M support, among others. It uses convention over configuration, which means it is pretty easy to get your first website running. Read more about it at `https://pagic.org/`.

Webview_deno

With many of the applications people use nowadays being written in JavaScript and living inside a web view, it was just a matter of time until they arrived on Deno. This module includes a Deno plugin and thus is still considered unstable. However, even though it has limitations and is an ongoing project, it already provides many of the interesting functionalities provided by Electron (the Node.js alternative).

On top of all the aforementioned packages, all the packages mentioned in *Chapter 4, Building a Web Application*, are worth having a look at. They're web frameworks that are evolving fast and, as we explored before, offer different benefits to the developers using them. If you are developing web applications with Deno, make sure you keep an eye on them. Check out their GitHub page at `https://github.com/webview/webview_deno`.

Do you think there is still functionality missing on Deno? Have you developed anything that you think would be useful to more people? The core of open source relies on those interesting pieces of software and the people behind them.

Made something you want to share? No worries—we've got you covered. In the next section, you'll learn how can you do it!

Publishing a package to Deno's official registry

Open source is, at its core, made of people and companies that use free software and have the desire to give back. When you create a piece of code that you think is interesting enough, you most likely want to share it. This is not only a way of helping other people but also a way to improve your own code.

Open source and this culture of sharing is what made Deno, Node.js, and many other technologies you probably use a reality. Since this book is all about Deno, it wouldn't make sense to finish it without going over this topic.

Deno has an official module registry that we've used before. This is a place where anyone with a GitHub account can share their own modules with the community, and it provides automation and caching mechanisms to keep different versions of modules.

What we're going to do next is publish a module of our own to this same registry.

We'll use a piece of software that, until now, we made available via the direct link to GitHub. This works but it has neither clear versioning nor any type of cache, making it unusable if the code is deleted from GitHub.

Remember when we used an `AuthRepository` that lived inside a package we called `jwt-auth`? At the time, for practical reasons, we used a direct GitHub link, but from now on we'll publish it in Deno's module registry.

We'll use the exact same code that's hosted on GitHub but publish it with the name of `deno_web_development_jwt_auth`. We're choosing this name to make it very clear that it is part of this book's journey. We also don't want to grab meaningful names from the registry for packages developed for learning purposes.

Let's go! Proceed as follows:

1. Create a repository for the module you want to publish. As mentioned, we'll be using the `jwt-auth` module from *Chapter 6*, *Adding Authentication and Connecting to the Database* (`https://github.com/PacktPublishing/Deno-Web-Development/tree/master/Chapter06/jwt-auth`), but feel free to use any other module of your choice.

2. Clone the recently created `git` repository by following GitHub's instructions. Make sure you copy your module's files to this repository folder, and run the following commands (these are the same as those presented in GitHub's instructions):

```
$ echo "# <Name of your package>" >> README.md
$ git init
$ git add .
$ git commit -m "first commit"
$ git branch -M main
$ git remote add origin git@github.com:<your-username>/
<your_package_name>.git
$ git push -u origin main
```

3. Go to https://deno.land/x and click the **Add a module** button (you might need to scroll a little to find it), as illustrated in the following screenshot:

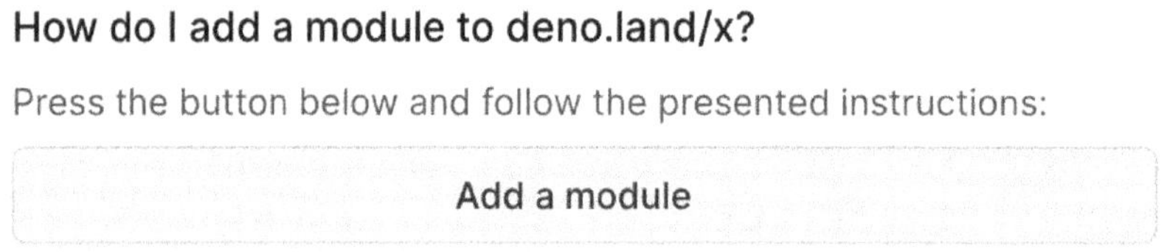

How do I add a module to deno.land/x?

Press the button below and follow the presented instructions:

> Add a module

Figure 10.2 – The Add a module button in the Deno module registry

4. Input the name of the module in the box that appears and click **Next**.

We'll be using deno_web_development_jwt_auth as the name of the package, but for obvious reasons you can't do the same.

Keep in mind that you should use a testing name if you're publishing a module for testing reasons. We don't want to be using "real" module names with modules used for testing purposes.

5. In the next box that appears, choose the directory where the code to be published is living.

For our module, which will contain the jwt-auth code from *Chapter 6, Adding Authentication and Connecting to the Database*, we'll leave this blank because it is living on the root of the (new) repository created in *Step 1*.

6. Now, it's just a matter of adding the webhook by following the instructions.

The Deno module registry uses GitHub webhooks to get updates for a package. These webhooks should be triggered by new branches or tags, and Deno's module registry will then create a version out of those GitHub tags.

The next instructions are presented on Deno's page, but I'll list them here for practical reasons:

a. Navigate to the repository you want to add on GitHub.

b. Go to the **Settings** tab.

c. Click on the **Webhooks** tab.

d. Click on the **Add webhook** button.

e. Enter the following **Uniform Resource Locator** (**URL**) in the **Payload URL** field: `https://api.deno.land/webhook/gh/<package_name>` (the package name should be the same as the one you chose in *Step 4*).

f. Select `application/json` as the content type.

g. Select **Let me select individual events**.

h. Select only the **Branch or tag creation** event.

i. Press **Add webhook**.

7. Now, it is just a matter of creating a release, which, as we mentioned, is done via `git` tags. Assuming you have already committed your package's code in *Step 2*, we just need to create and push this tag, as follows:

```
$ git tag v0.0.1
$ git push origin --tags
Enumerating objects: 5, done.
Counting objects: 100% (5/5), done.
Delta compression using up to 8 threads
Compressing objects: 100% (3/3), done.
Writing objects: 100% (3/3), 748 bytes | 748.00 KiB/s,
done.
Total 3 (delta 1), reused 0 (delta 0)
remote: Resolving deltas: 100% (1/1), completed with 1
local object.
To github.com:asantos00/deno_web_development_jwt_auth.git
 * [new tag]         v0.0.1 -> v0.0.1
```

8. If we now navigate to `https://deno.land/x` and search for the name of your package (`deno_web_development_jwt_auth`, in our example), it should appear there, as you can see in the following screenshot:

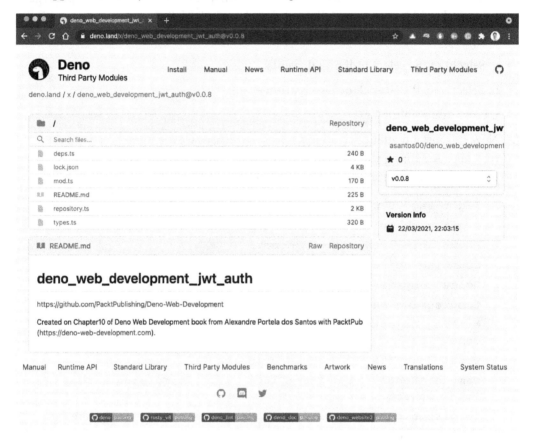

Figure 10.3 – A published package on Deno's module registry

And that is it—that is all you need to start sharing your amazing Deno code with the community! From now on, you can not only use Deno to build applications but also create packages and give something back to the community.

This wraps it up for this section and for the book—thanks for keeping up with it and reaching the end. We hope it was useful for you and that it helped you in learning Deno, and also that you are as excited about it as we are.

I'd be very happy to get in touch if you think there's anything I can help on. Feel free to reach me by the contacts present on the book's preface, via GitHub or Twitter.

Summary

First and foremost, thanks to all of you who stayed until the end of the book! I hope it was an interesting journey for you, one that fulfilled your expectations and addressed many of your questions and concerns about Deno.

This is only the start of a (hopefully big) journey. Deno is growing and you're now part of it. The more you use it and contribute back, the more it will get better. If, like me, you believe it offers a lot of benefits that can make it a game changer for writing JavaScript applications, do not wait to share your enthusiasm.

Lots of people like us are helping to push Deno forward, helping communities, developing modules, and opening pull requests. At the end of the day, using it for projects where it is well suited is the best recommendation you can make.

Throughout the book, I not only tried to highlight Deno's advantages but also tried to make it very clear that it is not, and will not be, a silver bullet. It offers a great set of advantages, especially when compared to Node.js, in the same set of use cases (as you can check in *Chapter 1*, *What is Deno?*). As we addressed in this chapter, there are many features being added that will enable Deno to be used for more and more use cases, but I am sure that there's a lot to come that we don't even know about.

From here onward, it is all up to you. I hope this book left you excited and that you can't wait to write Deno applications.

The next best step is to write applications yourself. This will lead you to research, talk with people, and solve your own problems. I tried to ease your path forward as much as possible by addressing some of the most common concerns.

I am sure that there are lots of online resources, articles, and books, but the true place to get better with Deno is still the Discord channel and the GitHub repositories. Those are the places where the news comes in firsthand!

I can't wait to see what you'll build next.

`Packt.com`

Subscribe to our online digital library for full access to over 7,000 books and videos, as well as industry leading tools to help you plan your personal development and advance your career. For more information, please visit our website.

Why subscribe?

- Spend less time learning and more time coding with practical eBooks and Videos from over 4,000 industry professionals

- Improve your learning with Skill Plans built especially for you

- Get a free eBook or video every month

- Fully searchable for easy access to vital information

- Copy and paste, print, and bookmark content

Did you know that Packt offers eBook versions of every book published, with PDF and ePub files available? You can upgrade to the eBook version at `packt.com` and as a print book customer, you are entitled to a discount on the eBook copy. Get in touch with us at `customercare@packtpub.com` for more details.

At `www.packt.com`, you can also read a collection of free technical articles, sign up for a range of free newsletters, and receive exclusive discounts and offers on Packt books and eBooks.

Other Books You May Enjoy

If you enjoyed this book, you may be interested in these other books by Packt:

Full-Stack React, TypeScript, and Node

David Choi

ISBN: 978-1-83921-993-1

- Discover TypeScript's most important features and how they can be used to improve code quality and maintainability
- Understand what React Hooks are and how to build React apps using them
- Implement state management for your React app using Redux
- Set up an Express project with TypeScript and GraphQL from scratch
- Build a fully functional online forum app using React and GraphQL
- Add authentication to your web app using Redis
- Save and retrieve data from a Postgres database using TypeORM
- Configure NGINX on the AWS cloud to deploy and serve your apps

Node Cookbook - Fourth Edition

Bethany Griggs

ISBN: 978-1-83855-875-8

- Understand the Node.js asynchronous programming model
- Create simple Node.js applications using modules and web frameworks
- Develop simple web applications using web frameworks such as Fastify and Express
- Discover tips for testing, optimizing, and securing your web applications
- Create and deploy Node.js microservices
- Debug and diagnose issues in your Node.js applications

Packt is searching for authors like you

If you're interested in becoming an author for Packt, please visit `authors.packtpub.com` and apply today. We have worked with thousands of developers and tech professionals, just like you, to help them share their insight with the global tech community. You can make a general application, apply for a specific hot topic that we are recruiting an author for, or submit your own idea.

Leave a review - let other readers know what you think

Please share your thoughts on this book with others by leaving a review on the site that you bought it from. If you purchased the book from Amazon, please leave us an honest review on this book's Amazon page. This is vital so that other potential readers can see and use your unbiased opinion to make purchasing decisions, we can understand what our customers think about our products, and our authors can see your feedback on the title that they have worked with Packt to create. It will only take a few minutes of your time, but is valuable to other potential customers, our authors, and Packt. Thank you!

Index

www.ingramcontent.com/pod-product-compliance
Lightning Source LLC
Chambersburg PA
CBHW062109050326
40690CB00016B/3257